NIO高并发WebSocket框架开发

微课视频版

刘宁萌 ◎ 编著

清華大学出版社
北京

内 容 简 介

本书全面系统地介绍了Java网络套接字相关的核心知识点，把官方BIO、NIO的核心知识点用链路的方式讲解清楚。读者在学习的过程中需要循序渐进，核心知识点的理解很重要。希望读者在此过程中能够成长和拥有自己的思维。

全书共18章，第1章讲解多线程并发经典案例，第2～5章介绍IO字节、字符输入输出流，第6章介绍框架基石反射机制，第7章介绍类加载器，第8章介绍网络基础知识，第9章介绍BIO模式下的Socket套接字，第10章介绍NIO模式下的缓冲区，第11章介绍NIO模式下的字符编解码，第12章介绍NIO模式下的Socket套接字通道，第13章介绍泛型声明，第14章介绍List集合框架，第15章介绍Set集合框架，第16章介绍Map集合框架，第17章介绍开源WebSocket框架实战，第18章介绍JDK 21中的虚拟线程。

本书每个章节都有相对应的代码验证环节，提供了大量应用实例。

本书适合有一定Java基础知识和网络通信基础的读者阅读。本书可作为想自研NIO套接字框架、自定义协议、网络通信服务相关的软件开发人员的参考用书。

图书在版编目（CIP）数据

NIO高并发WebSocket框架开发：微课视频版 / 刘宁萌编著. -- 北京：清华大学出版社，2025. 1. （计算机技术开发与应用丛书）. -- ISBN 978-7-302-67912-7

Ⅰ. TP312.8

中国国家版本馆CIP数据核字第20252LT107号

责任编辑：赵佳霓
封面设计：吴 刚
责任校对：郝美丽
责任印制：宋 林

出版发行：清华大学出版社
网 址：https://www.tup.com.cn，https://www.wqxuetang.com
地 址：北京清华大学学研大厦A座 **邮 编**：100084
社 总 机：010-83470000 **邮 购**：010-62786544
投稿与读者服务：010-62776969，c-service@tup.tsinghua.edu.cn
质量反馈：010-62772015，zhiliang@tup.tsinghua.edu.cn
课件下载：https://www.tup.com.cn，010-83470236
印 装 者：北京鑫海金澳胶印有限公司
经 销：全国新华书店
开 本：186mm×240mm **印 张**：21.5 **字 数**：483千字
版 次：2025年3月第1版 **印 次**：2025年3月第1次印刷
印 数：1～1500
定 价：89.00元

产品编号：105241-01

前言

PREFACE

Java 在 1.4 版本之前使用的是 BIO 模型。此模型是阻塞式模型，一个套接字连接需要一个线程来维护，一台计算机能开辟的线程数量有限，并且过多的线程会导致频繁的上下文切换，所以 BIO 的并发性能较低。

Java 在 1.4 版本推出了 NIO 模型。此模型是多路复用的非阻塞式模型，配合 Selector、SelectableChannel、SelectionKey 可以搭建多路复用的非阻塞式套接字网络服务。

Java 在 1.7 版本推出了 AIO 模型。此模型是在 NIO 的基础上扩展了异步回调。可以理解为封装了一个多线程异步回调的框架，任何框架的使用都有两面性，一方面是封装好的服务只需调用就可以了；另一方面也会受到框架设计使用等方面的制约。

Java 在 21 版本推出了虚拟线程。虚拟线程可以创建百万级虚拟线程，但不能创建百万级平台线程。虚拟线程的切换成本较低，平台线程的切换成本较高。

本书特点

本书以 JDK 17 长期维护版本为示例，详细地介绍了 Java 套接字网络通信相关的知识体系，让读者不仅知其然，而且知其所以然。本书对套接字网络通信相关的理论分门别类，层层递进地进行详细叙述和透彻分析，既体现了各知识点之间的联系，又兼顾了其渐进性。本书在介绍每个知识点时都给出了该知识点的应用场景，同时配合代码演示，让读者更能明白其实现原理和使用方式。本书丰富的免费配套资源包括源代码、完整的配套视频。

本书主要内容

第 1 章介绍了单例双重检查锁机制、happens-before 原则、LockSupport 底层锁。

第 2 章介绍了数据流、序列化、文件系统提供的字节输入流功能。

第 3 章介绍了数据流、序列化、文件系统提供的字节输出流功能。

第 4 章介绍了数据流、序列化、文件系统提供的字符输入流功能。

第 5 章介绍了数据流、序列化、文件系统提供的字符输出流功能。

第 6 章介绍了 Class 类框架基石反射机制。

第 7 章介绍了 ClassLoader 类加载器，并实现自定义的类加载器。

第 8 章介绍了网络基础知识，如 IP 地址、网络接口、UDP。

第 9 章介绍了 BIO 模式下的 Socket 网络套接字服务。

第 10 章介绍了 NIO 模式下的 Buffer 缓冲区。

第 11 章介绍了 NIO 模式下的字符集、解码器、编码器。

第 12 章介绍了 NIO 模式下搭建多路复用的非阻塞式套接字网络服务。

第 13 章介绍了泛型的使用。

第 14 章介绍了 List 集合框架。

第 15 章介绍了 Set 集合框架。

第 16 章介绍了 Map 集合框架。

第 17 章介绍了自研 WebSocket 框架。基于 Reactor 设计模式，使用原生 NIO 实现的 WebSocket 网络框架。支持多线程、高并发、TLS 安全层协议。

第 18 章介绍了 JDK 21 中的虚拟线程。

资源下载提示

素材(源码)等资源：扫描目录上方的二维码下载。

视频等资源：扫描封底的文泉云盘防盗码，再扫描书中相应章节的二维码，可以在线学习。

读者对象

本书适合想自研 NIO 套接字框架、自定义协议、网络通信服务的 Java 开发人员阅读。

致谢

特别感谢清华大学出版社赵佳霓编辑。感谢她对本书专业且高效的审阅、对书中各种表意方式和文笔的润色建议，以及推动本书的出版。同时也感谢参与本书出版的所有出版社的老师，在大家的辛勤努力下，才有了本书的顺利出版。

刘宁萌

2024 年 10 月

目录

CONTENTS

本书源码

第1章

CHAPTER 1

并发扩展

本章介绍在多线程并发场景下的经典案例，例如单例双重检查锁机制、happens-before原则、LockSupport底层锁。

1.1 单例双重检查锁

单例双重检查锁机制是一个非常经典的案例，通过此案例有助于读者理解多线程并发的核心概念，并理解重排序可能对代码执行产生的影响。

1.1.1 第一阶段

最基础的一个单例模式，多线程并发单例失败，并且可能存在重排序问题，代码如下：

```
//第1章/one/DoubleCheckSingleton.java
public class DoubleCheckSingleton {
    private DoubleCheckSingleton(){}

    private static DoubleCheckSingleton doubleCheckSingleton;

    public static void main(String[] args) {
        System.out.println(callOne());
        System.out.println(callOne());
        System.out.println(callOne());
    }

    //多线程并发单例失败，并且可能存在重排序问题
    public static DoubleCheckSingleton callOne(){
        if(doubleCheckSingleton == null){
            doubleCheckSingleton = new DoubleCheckSingleton();
        }
        return doubleCheckSingleton;
    }
}
```

执行结果如下：

```
cn.kungreat.niobook.one.one.DoubleCheckSingleton@776ec8df
cn.kungreat.niobook.one.one.DoubleCheckSingleton@776ec8df
cn.kungreat.niobook.one.one.DoubleCheckSingleton@776ec8df
```

1.1.2 第二阶段

基于第一阶段的单例升级模式，多线程并发单例失败，并且可能存在重排序问题，代码如下：

```
//第1章/one/DoubleCheckSingleton.java
public class DoubleCheckSingleton {
    private DoubleCheckSingleton(){}

    private static DoubleCheckSingleton doubleCheckSingleton;

    public static void main(String[] args) {
        System.out.println(callTwo());
        System.out.println(callTwo());
        System.out.println(callTwo());
    }

    //多线程并发单例失败,并且可能存在重排序问题
    public static DoubleCheckSingleton callTwo(){
        if(doubleCheckSingleton == null){
            synchronized (DoubleCheckSingleton.class){
                doubleCheckSingleton = new DoubleCheckSingleton();
            }
        }
        return doubleCheckSingleton;
    }
}
```

执行结果如下：

```
cn.kungreat.niobook.one.one.DoubleCheckSingleton@776ec8df
cn.kungreat.niobook.one.one.DoubleCheckSingleton@776ec8df
cn.kungreat.niobook.one.one.DoubleCheckSingleton@776ec8df
```

注意：加了 synchronized 锁后防止了多线程并发，但是仍然可能存在重复创建对象的情况。

1.1.3 第三阶段

基于第二阶段的单例升级模式，多线程并发单例成功，但是可能存在重排序问题，代码如下：

```
//第 1 章/one/DoubleCheckSingleton.java
public class DoubleCheckSingleton {
    private DoubleCheckSingleton(){}

    private static DoubleCheckSingleton doubleCheckSingleton;

    public static void main(String[] args) {
        System.out.println(callThree());
        System.out.println(callThree());
        System.out.println(callThree());
    }

    //多线程并发单例成功,但是可能存在重排序问题
    public static DoubleCheckSingleton callThree(){
        if(doubleCheckSingleton == null){
            synchronized (DoubleCheckSingleton.class){
                if(doubleCheckSingleton == null){
                    doubleCheckSingleton = new DoubleCheckSingleton();
                }
            }
        }
        return doubleCheckSingleton;
    }

}
```

执行结果如下：

```
cn.kungreat.niobook.one.one.DoubleCheckSingleton@776ec8df
cn.kungreat.niobook.one.one.DoubleCheckSingleton@776ec8df
cn.kungreat.niobook.one.one.DoubleCheckSingleton@776ec8df
```

注意：此单例模式是成功的,但是可能存在多线程并发重排序问题。

1.1.4 最终阶段

基于第三阶段的单例升级模式,多线程并发单例成功,并且解决了重排序问题,代码如下：

```
//第 1 章/one/DoubleCheckSingleton.java
public class DoubleCheckSingleton {
    private DoubleCheckSingleton(){}
    /* volatile 解决可见性、重排序问题 */
    private static volatile DoubleCheckSingleton doubleCheckSingleton;

    public static void main(String[] args) {
        System.out.println(callThree());
```

```
        System.out.println(callThree());
        System.out.println(callThree());
    }

    //多线程并发单例成功
    public static DoubleCheckSingleton callThree(){
        if(doubleCheckSingleton == null){
            synchronized (DoubleCheckSingleton.class){
                if(doubleCheckSingleton == null){
                    doubleCheckSingleton = new DoubleCheckSingleton();
                }
            }
        }
        return doubleCheckSingleton;
    }

}
```

执行结果如下：

```
cn.kungreat.niobook.one.one.DoubleCheckSingleton@776ec8df
cn.kungreat.niobook.one.one.DoubleCheckSingleton@776ec8df
cn.kungreat.niobook.one.one.DoubleCheckSingleton@776ec8df
```

注意：可见性、重排序并不影响单个线程的执行结果，只在多线程并发时可能产生异常。

1.1.5 扩展 happens-before

思考以下代码是否会因为重排序而造成异常，代码如下：

```
//第1章/one/DoubleCheckSingleton.java
public class DoubleCheckSingleton {

    private DoubleCheckSingleton(){}
    private static DoubleCheckSingleton doubleCheckSingleton;

    public static void main(String[] args) {
        System.out.println(callFour());
        System.out.println(callFour());
        System.out.println(callFour());
    }

    public static DoubleCheckSingleton callFour(){
        synchronized (DoubleCheckSingleton.class){
            if(doubleCheckSingleton == null){
                doubleCheckSingleton = new DoubleCheckSingleton();
```

```
            }
        }
        return doubleCheckSingleton;
    }
}
```

注意：以上单例模式多线程并发单例成功，重排序问题不会影响最终的执行结果，但是方法的执行效率低。

1.2　LockSupport 锁

此类提供了线程的阻塞和唤醒功能，属于基本的锁功能实现。使用此类需要自己设计一套锁功能机制。官方的抽象队列同步器(Abstract Queued Synchronizer，AQS)也是基于此类设计的。

本节基于官方文档介绍常用方法，核心方法配合代码示例以方便读者理解。

1. park()

使当前执行线程阻塞等待，如果没有被唤醒，则一直阻塞等待，代码如下：

```
//第 1 章/two/LockSupportTest.java
public class LockSupportTest {

    public static void main(String[] args) throws InterruptedException {
        Thread thread = new Thread(new Runnable() {
            @Override
            public void run() {
                System.out.println("start");
                LockSupport.park();
                System.out.println("end");
            }
        },"A");
        thread.start();
        Thread.sleep(300);
        System.out.println("main - end");
    }

}
```

执行结果如下：

```
start
main - end
WAITING
```

注意：线程A一直处于阻塞等待中并且Java虚拟机(Java Virtual Machine,JVM)也没有关闭。

2. park(Object blocker)

使当前执行线程阻塞等待，如果没有被唤醒，则一直阻塞等待，并设置此线程停放的同步对象。接收Object入参，作为此线程停放的同步对象，代码如下：

```
//第1章/two/LockSupportTest.java
public class LockSupportTest {

    public static void main(String[] args) throws InterruptedException {
        Thread thread = new Thread(new Runnable() {
            @Override
            public void run() {
                System.out.println("start");
                LockSupport.park("block - test");
                System.out.println("end");
            }
        },"A");
        thread.start();
        Thread.sleep(300);
        System.out.println("main - end");
        System.out.println(LockSupport.getBlocker(thread));
        System.out.println(thread.getState());
    }

}
```

执行结果如下：

```
start
main - end
block - test
WAITING
```

3. parkNanos(long nanos)

使当前执行线程阻塞等待，直到它被唤醒或者超过最长等待时间。接收long入参，作为最长等待时间的纳秒数，代码如下：

```
//第1章/two/LockSupportTest.java
public class LockSupportTest {

    public static void main(String[] args) throws InterruptedException {
        Thread thread = new Thread(new Runnable() {
            @Override
            public void run() {
                System.out.println("start");
```

```
                LockSupport.parkNanos(3000000000L);
                System.out.println("end");
            }
        },"A");
        thread.start();
        Thread.sleep(300);
        System.out.println("main - end");
        System.out.println(thread.getState());
    }

}
```

执行结果如下：

```
start
main - end
TIMED_WAITING
End
```

4. parkNanos(Object blocker, long nanos)

使当前执行线程阻塞等待，直到它被唤醒或者超过最长等待时间，并设置此线程停放的同步对象。接收 Object 入参，作为此线程停放的同步对象，接收 long 入参，作为最长等待时间的纳秒数，代码如下：

```
//第 1 章/two/LockSupportTest.java
public class LockSupportTest {

    public static void main(String[] args) throws InterruptedException {
        Thread thread = new Thread(new Runnable() {
            @Override
            public void run() {
                System.out.println("start");
                LockSupport.parkNanos("block - test",3000000000L);
                System.out.println("end");
            }
        },"A");
        thread.start();
        Thread.sleep(300);
        System.out.println("main - end");
        System.out.println(LockSupport.getBlocker(thread));
        System.out.println(thread.getState());
    }

}
```

执行结果如下：

```
start
main - end
```

```
block - test
TIMED_WAITING
end
```

5. parkUntil(long deadline)

使当前执行线程阻塞等待，直到它被唤醒或者到达指定截止时间。接收 long 入参，作为指定截止时间的毫秒数，代码如下：

```
//第 1 章/two/LockSupportTest.java
public class LockSupportTest {

    public static void main(String[] args) throws InterruptedException {
        Thread thread = new Thread(new Runnable() {
            @Override
            public void run() {
                System.out.println("start");
                LockSupport.parkUntil(System.currentTimeMillis() + 3000);
                System.out.println("end");
            }
        },"A");
        thread.start();
        Thread.sleep(300);
        System.out.println("main - end");
        System.out.println(LockSupport.getBlocker(thread));
        System.out.println(thread.getState());
    }

}
```

执行结果如下：

```
start
main - end
null
TIMED_WAITING
end
```

6. parkUntil(Object blocker, long deadline)

使当前执行线程阻塞等待，直到它被唤醒或者到达指定截止时间，并设置此线程停放的同步对象。接收 Object 入参，作为此线程停放的同步对象，接收 long 入参，作为指定截止时间的毫秒数。

7. unpark(Thread thread)

唤醒指定的线程（如果它正在阻塞等待中），代码如下：

```
//第 1 章/two/LockSupportTest.java
public class LockSupportTest {
```

```
    public static void main(String[] args) throws InterruptedException {
        Thread thread = new Thread(new Runnable() {
            @Override
            public void run() {
                System.out.println("start");
                LockSupport.park();
                System.out.println("end");
            }
        },"A");
        thread.start();
        Thread.sleep(300);
        System.out.println("main - end");
        System.out.println(LockSupport.getBlocker(thread));
        System.out.println(thread.getState());
        LockSupport.unpark(thread);
    }

}
```

执行结果如下：

```
start
main - end
null
WAITING
end
```

8．setCurrentBlocker(Object blocker)

设置当前执行线程停放的同步对象。接收 Object 入参，作为当前执行线程停放的同步对象。

9．getBlocker(Thread t)

获取指定线程停放的同步对象，接收 Thread 入参，作为指定的线程，代码如下：

```
//第 1 章/two/LockSupportTest.java
public class LockSupportTest {

    public static void main(String[] args) {
        LockSupport.setCurrentBlocker("block - main");
        System.out.println(LockSupport.
                    getBlocker(Thread.currentThread()));
    }

}
```

执行结果如下：

```
block - main
```

小结

通过本章的学习，读者应更深入地理解单线程、多线程并发程序之间的区别。

习题

1. 判断题

（1）关键字 volatile 可以解决可见性和重排序问题。（　　）

（2）单线程执行的程序不受可见性和重排序的影响。（　　）

（3）单例双重检查锁机制需要使用 volatile 关键字。（　　）

2. 选择题

（1）LockSupport 类可以使当前执行线程阻塞等待的方法有（　　）。（多选）

A. park()　　B. park(Object blocker)

C. parkUntil(long deadline)　　D. unpark(Thread thread)

（2）LockSupport 类可以唤醒当前执行线程的方法是（　　）。（单选）

A. park()　　B. park(Object blocker)

C. parkUntil(long deadline)　　D. unpark(Thread thread)

3. 填空题

（1）查看执行结果并补充代码，代码如下：

```
//第 1 章/answer/LockQuestion.java
public class LockQuestion {
    public static void main(String[ ] args) throws Exception {
        Thread thread = new Thread(new Runnable() {
            @Override
            public void run() {
                System.out.println("start");
                LockSupport.park("block - test");
                System.out.println("end");
            }
        },"A");
        thread.start();
        LockSupport.______________________________;
        Thread.sleep(300);
        System.out.println(LockSupport.getBlocker(thread));
        System.out.println("main - end");
        System.out.println(LockSupport.
            getBlocker(Thread.currentThread()));
        LockSupport.______________________________;
    }
}
```

执行结果如下：

```
start
block - test
main - end
block - 2
end
```

(2) 完成单例双重检查锁机制，代码如下：

```
//第 1 章/answer/DoubleCheckSingleton.java
public class DoubleCheckSingleton {

    private DoubleCheckSingleton(){}

    private static volatile DoubleCheckSingleton doubleCheckSingleton;

    public static DoubleCheckSingleton checkSingleton (){
        if(_________________________________){
            synchronized (DoubleCheckSingleton.class){
                if(_________________________________){
                    doubleCheckSingleton = new DoubleCheckSingleton();
                }
            }
        }
        return doubleCheckSingleton;
    }

}
```

第 2 章

CHAPTER 2

IO 字节输入流

IO 字节输入流通过数据流、序列化、文件系统提供字节输入功能，输入和输出是相对的概念。

77min

2.1 File 类

此类表示文件系统，提供了目录和文件的相关描述信息。

2.1.1 构造器

File 构造器见表 2-1。

表 2-1 File 构造器

构 造 器	描 述
File(File parent, String child)	构造新的对象，指定上级文件系统，指定子文件系统名称
File(String pathname)	构造新的对象，指定文件系统全路径名称
File(String parent, String child)	构造新的对象，指定上级文件系统名称，指定子文件系统名称
File(URI uri)	构造新的对象，指定统一资源标识

2.1.2 常用方法

本节基于官方文档介绍常用方法，核心方法配合代码示例以方便读者理解。

1. canExecute()

检查当前文件系统是否有可执行权限，返回 boolean 值。

2. canRead()

检查当前文件系统是否有可读取权限，返回 boolean 值。

3. canWrite()

检查当前文件系统是否有可写入权限，返回 boolean 值。

4. compareTo(File pathname)

比较当前文件系统和指定的文件系统是否相同,返回 int 值。接收 File 入参,作为指定的文件系统,代码如下:

```
//第 2 章/one/FileTest.java
public class FileTest {
    public static void main(String[ ] args) throws IOException {
        //系统路径分隔符
        System.out.println(File.pathSeparator);
        System.out.println(File.pathSeparatorChar);
        //系统文件分隔符
        System.out.println(File.separator);
        System.out.println(File.separatorChar);
        File file = new File("D:\\temp\\src\\test.txt");
        System.out.println(file.
                    compareTo(new File("D:\\temp\\src\\test.txt")));
    }
}
```

执行结果如下:

```
;
;
\
\
0
```

5. createNewFile()

如果当前文件不存在,则创建它,返回 boolean 值,代码如下:

```
//第 2 章/one/FileTest.java
public class FileTest {
public static void main(String[ ] args) throws IOException {
    //文件默认不存在
        File file = new File("D:\\temp\\src\\test9.txt");
        System.out.println(file.createNewFile());
        System.out.println(file.createNewFile());
    }
}
```

执行结果如下:

```
true
false
```

6. createTempFile(String prefix, String suffix)

静态方法,在默认临时文件目录中创建空文件,使用指定的前缀和后缀生成其名称。接收 String 入参,作为指定的前缀,接收 String 入参,作为指定的后缀,代码如下:

```
//第 2 章/one/FileTest.java
public class FileTest {
    public static void main(String[] args) throws IOException {
        System.out.println(File.createTempFile("test", ".txt"));
    }
}
```

执行结果如下：

```
C:\Users\mydre\AppData\Local\Temp\test14465305846336034682.txt
```

7. delete()

删除当前文件系统，返回 boolean 值，代码如下：

```
//第 2 章/one/FileTest.java
public class FileTest {
    public static void main(String[] args) throws IOException {
        //这个文件默认存在
        File file = new File("D:\\temp\\src\\test9.txt");
        System.out.println(file.delete());
        System.out.println(file.delete());
    }
}
```

执行结果如下：

```
true
false
```

8. deleteOnExit()

当虚拟机终止时删除当前文件系统。

9. equals(Object obj)

比较当前文件系统和指定的对象，返回 boolean 值。接收 Object 入参，作为指定的对象。

10. exists()

检查当前文件系统是否存在，返回 boolean 值。

11. getAbsolutePath()

返回当前文件系统的绝对路径名称。

12. getCanonicalPath()

返回当前文件系统的规范路径名称。

13. getFreeSpace()

返回当前文件系统所在分区中未分配的字节数，代码如下：

```
//第 2 章/one/FileTest.java
public class FileTest {
    public static void main(String[] args) throws IOException {
        File file = new File("D:\\temp\\src\\test.txt");
        System.out.println(file.getFreeSpace());
    }
}
```

执行结果如下：

```
60840787968
```

在 Windows 系统下的分区概念，如图 2-1 所示。

图 2-1　Windows 系统下的分区概念

14. getName()

返回当前文件系统所表示的名称，代码如下：

```
//第 2 章/one/FileTest.java
public class FileTest {
    public static void main(String[] args) throws IOException {
        File file = new File("D:\\temp\\src\\test.txt");
        System.out.println(file.getName());
        File file1 = new File("D:\\temp\\src");
        System.out.println(file1.getName());
    }
}
```

执行结果如下：

```
test.txt
src
```

15. getParent()

返回当前文件系统的上级，如果不存在上级，则返回空。

16. getPath()

返回当前文件系统的全路径名称，代码如下：

```
//第 2 章/one/FileTest.java
public class FileTest {
    public static void main(String[] args) throws IOException {
        File file = new File("D:\\temp\\src\\test.txt");
        System.out.println(file.getPath());
        File file1 = new File("D:\\temp\\src6");
        System.out.println(file1.getPath());
    }
}
```

执行结果如下：

```
D:\temp\src\test.txt
D:\temp\src6
```

17. getTotalSpace()

返回当前文件系统所在分区中总的字节数。

18. isAbsolute()

检查当前文件系统是否是绝对路径，返回 boolean 值。

19. isDirectory()

检查当前文件系统是否是目录，返回 boolean 值。

20. isFile()

检查当前文件系统是否是文件，返回 boolean 值。

21. lastModified()

返回当前文件系统最后修改时间的时间戳。

22. length()

返回当前文件系统所表示数据的长度，代码如下：

```
//第 2 章/one/FileTest.java
public class FileTest {
    public static void main(String[] args) throws IOException {
```

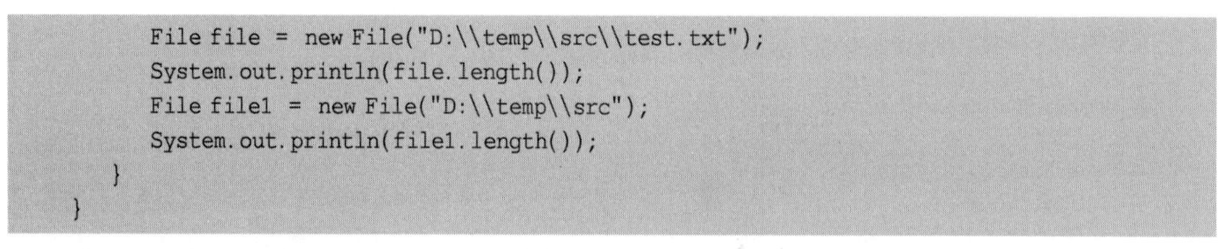

```
        File file = new File("D:\\temp\\src\\test.txt");
        System.out.println(file.length());
        File file1 = new File("D:\\temp\\src");
        System.out.println(file1.length());
    }
}
```

执行结果如下：

```
9
0
```

23. list()

返回当前文件系统下的子文件系统名称，代码如下：

```
//第 2 章/one/FileTest.java
public class FileTest {
    public static void main(String[] args) throws IOException {
        File file = new File("D:\\temp\\src\\test.txt");
        System.out.println(Arrays.toString(file.list()));
        File file1 = new File("D:\\temp\\src");
        System.out.println(Arrays.toString(file1.list()));
    }
}
```

执行结果如下：

```
null
[main, test.txt, test2.txt, test5.txt]
```

24. listFiles()

返回当前文件系统下的子文件系统，代码如下：

```
//第 2 章/one/FileTest.java
public class FileTest {
    public static void main(String[] args) throws IOException {
        File file = new File("D:\\temp\\src\\test.txt");
        System.out.println(Arrays.toString(file.listFiles()));
        File file1 = new File("D:\\temp\\src");
        System.out.println(Arrays.toString(file1.listFiles()));
    }
}
```

执行结果如下：

```
null
[D:\temp\src\main, D:\temp\src\test.txt, D:\temp\src\test2.txt, D:\temp\src\test5.txt]
```

25. mkdir()

创建当前文件系统所表示的目录，返回 boolean 值，代码如下：

```
//第 2 章/one/FileTest.java
public class FileTest {
    public static void main(String[] args) throws IOException {
        File file = new File("D:\\temp\\src\\test666.txt");
        System.out.println(file.mkdir());
        File file1 = new File("D:\\temp\\src\\temp");
        System.out.println(file1.mkdir());
    }
}
```

执行结果如下：

```
true
true
```

在 Windows 系统下产生的效果，如图 2-2 所示。

名称	修改日期	类型	大小
main	2023/11/1 13:17	文件夹	
temp	2023/11/2 16:08	文件夹	
test666.txt	2023/11/2 16:08	文件夹	
test.txt	2023/11/1 15:17	文本文档	1 KB
test2.txt	2021/10/18 22:24	文本文档	1 KB
test5.txt	2023/11/1 15:35	文本文档	1 KB

新建的目录

图 2-2 新建目录

26. mkdirs()

创建当前文件系统所表示的目录，包括必需的但不存在的上级目录，返回 boolean 值。

27. renameTo(File dest)

重命名当前文件系统，返回 boolean 值。

28. setExecutable(boolean executable)

设置所有者对当前文件系统的可执行权限，返回 boolean 值。接收 boolean 入参，作为可执行权限的逻辑值。

29. setReadable(boolean readable)

设置所有者对当前文件系统的可读取权限，返回 boolean 值。接收 boolean 入参，作为可读取权限的逻辑值。

30. setWritable(boolean writable)

设置所有者对当前文件系统的可写入权限，返回 boolean 值。接收 boolean 入参，作为可写入权限的逻辑值。

31. setLastModified(long time)

设置当前文件系统最后修改时间的时间戳，返回 boolean 值。接收 long 入参，作为最后修改时间的时间戳。

2.1.3　搜索工具类

运用所学的知识点，编写一个文件名搜索工具类，代码如下：

```
//第 2 章/one/FileUtils.java
public class FileUtils {

    public static void printChild(File fl) {
        if (fl != null && fl.exists() && fl.isDirectory()) {
            File[] fs = fl.listFiles();
            if (fs != null) {
                for (File temp : fs) {
                    System.out.println(temp);
                    if (temp.isDirectory()) {
                        printChild(temp);
                    }
                }
            }
        }
    }

    public static void searchNames(File fl, String name) {
        if (fl != null && fl.exists() && fl.isDirectory()) {
            File[] fs = fl.listFiles();
            if (fs != null) {
                for (File temp : fs) {
                    if (temp.getName().contains(name)) {
                        System.out.println(temp);
                    }
                    if (temp.isDirectory()) {
                        searchNames(temp, name);
                    }
                }
            }
        }
    }
}
```

查询当前系统中所有包含 docx 的目录和文件，代码如下：

```
//第 2 章/one/FileSearch.java
//import 第 2 章/one/FileUtils.java
public class FileSearch {
    public static void main(String[] args) {
        for (File file : File.listRoots()) {
            new Thread(() -> FileUtils.searchNames(file,"docx")).start();
        }
    }
}
```

9min

2.2 InputStream 抽象类

此抽象类是所有字节输入流的超类。本节基于官方文档介绍常用方法，核心方法配合代码示例以方便读者理解。

1. available()

返回从当前输入流中可读取的字节数。

2. close()

关闭当前输入流并释放与此流相关联的任何资源。

3. mark(int readlimit)

标记当前输入流中数据指针的位置。接收 int 入参，作为在标记位变为无效前可读取的最大字节数。

4. markSupported()

检查当前输入流是否支持标记和重置方法，返回 boolean 值。

5. read()

返回从当前输入流中读取的下一字节数据。

6. read(byte[] b)

从当前输入流中读取字节数据并将它们存储到指定的数组中，返回读取到的字节总数。如果到达此流的末尾，则为－1。接收 byte[]入参，作为指定的数组。

7. readAllBytes()

返回从当前输入流中读取到的所有剩余字节。

8. readNBytes(byte[] b, int off, int len)

从当前输入流中读取字节数据并将它们存储到指定的数组中，返回读取到的字节总数。接收 byte[]入参，作为指定的数组，接收 int 入参，作为数组的起始索引，接收 int 入参，作为要读取的最大字节长度。

9. readNBytes(int len)

返回从当前输入流中读取到的最大长度的字节数据。接收 int 入参，作为要读取的最大字节长度。

10. reset()

把当前输入流中的数据指针设置为以前标记的值。

11. skip(long n)

跳过当前输入流中指定长度的字节数据，返回已跳过的实际值。接收 long 入参，作为指定长度。

12. skipNBytes(long n)

强制跳过当前输入流中指定长度的字节数据,接收 long 入参,作为指定长度。

26min

2.3 FileInputStream 类

文件字节输入流从文件系统的文件中获取输入字节。

2.3.1 构造器

FileInputStream 构造器见表 2-2。

表 2-2 FileInputStream 构造器

构 造 器	描 述
FileInputStream(File file)	构造新的对象,指定文件系统
FileInputStream(FileDescriptor fdObj)	构造新的对象,指定文件描述符
FileInputStream(String name)	构造新的对象,指定文件系统路径

2.3.2 常用方法

本节基于官方文档介绍常用方法,核心方法配合代码示例以方便读者理解。

1. available()

返回从当前输入流中可读取的字节数,代码如下:

```
//第 2 章/three/FileInputStreamTest.java
public class FileInputStreamTest {

    public static void main(String[] args) {
        try (FileInputStream fileInputStream = new
                FileInputStream("D:\\temp\\src\\test.txt")) {
            System.out.println(fileInputStream.available());
        } catch (Exception exception) {
            System.out.println(exception);
        }
    }
}
```

执行结果如下:

```
18
```

文件内容如图 2-3 所示。

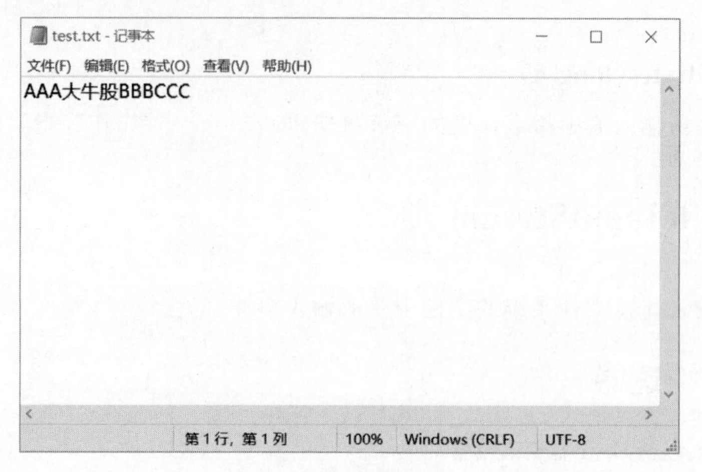

图 2-3 文件内容

2. close()

关闭当前输入流并释放与此流相关联的任何资源。

3. getFD()

返回表示与当前输入流正在使用的文件系统相关联的文件描述符。

4. markSupported()

检查当前输入流是否支持标记和重置方法，返回 boolean 值，代码如下：

```
//第 2 章/three/FileInputStreamTest.java
public class FileInputStreamTest {

    public static void main(String[] args) {
        try (FileInputStream fileInputStream = new FileInputStream("D:\\temp\\src\\test.
txt")) {
            System.out.println(fileInputStream.markSupported());
        } catch (Exception exception) {
            System.out.println(exception);
        }
    }
}
```

执行结果如下：

```
false
```

5. read()

返回从当前输入流中读取的下一字节数据，如果到达文件末尾，则为－1，代码如下：

```
//第 2 章/three/FileInputStreamTest.java
public class FileInputStreamTest {
```

```
    public static void main(String[ ] args) {
        try (FileInputStream fileInputStream = new
                    FileInputStream("D:\\temp\\src\\test.txt")) {
            System.out.println(fileInputStream.available());
            int bts = fileInputStream.read();
            while (bts != -1){
                System.out.println(bts);
                bts = fileInputStream.read();
            }
            System.out.println(fileInputStream.available());
        } catch (Exception exception) {
            System.out.println(exception);
        }
    }
}
```

执行结果如下：

```
18
65
65
65
229
164
167
231
137
155
232
130
161
66
66
66
67
67
67
0
```

6. read(byte[] b)

从当前输入流中读取字节数据并将它们存储到指定的数组中，返回读取到的字节总数。如果到达此流的末尾，则为－1。接收byte[]入参，作为指定的数组，代码如下：

```
//第 2 章/three/FileInputStreamTest.java
public class FileInputStreamTest {

    public static void main(String[ ] args) {
        try (FileInputStream fileInputStream = new
                FileInputStream("D:\\temp\\src\\test.txt")) {
```

```
            byte[] bts = new byte[fileInputStream.available()];
            System.out.println(fileInputStream.read(bts));
            System.out.println(new String(bts));
        } catch (Exception exception) {
            System.out.println(exception);
        }
    }
}
```

执行结果如下：

```
18
AAA 大牛股 BBBCCC
```

7. readAllBytes()

返回从当前输入流中读取到的所有剩余字节,代码如下：

```
//第 2 章/three/FileInputStreamTest.java
public class FileInputStreamTest {

    public static void main(String[] args) {
        try (FileInputStream fileInputStream = new
                    FileInputStream("D:\\temp\\src\\test.txt")) {
            byte[] bts = fileInputStream.readAllBytes();
            System.out.println(new String(bts));
        } catch (Exception exception) {
            System.out.println(exception);
        }
    }
}
```

执行结果如下：

```
AAA 大牛股 BBBCCC
```

8. readNBytes(byte[] b, int off, int len)

从当前输入流中读取字节数据并将它们存储到指定的数组中,返回读取到的字节总数。接收 byte[]入参,作为指定的数组,接收 int 入参,作为数组的起始索引,接收 int 入参,作为要读取的最大字节长度。

9. readNBytes(int len)

返回从当前输入流中读取到的最大长度的字节数据。接收 int 入参,作为要读取的最大字节长度。

10. skip(long n)

跳过当前输入流中指定长度的字节数据,返回已跳过的实际值。接收 long 入参,作为指定长度,代码如下：

```
//第 2 章/three/FileInputStreamTest.java
public class FileInputStreamTest {

    public static void main(String[] args) {
        try (FileInputStream fileInputStream = new
                        FileInputStream("D:\\temp\\src\\test.txt")) {
            System.out.println(fileInputStream.skip(3));
            byte[] bts = fileInputStream.readAllBytes();
            System.out.println(new String(bts));
        } catch (Exception exception) {
            System.out.println(exception);
        }
    }
}
```

执行结果如下：

```
3
大牛股 BBBCCC
```

11. skipNBytes(long n)

强制跳过当前输入流中指定长度的字节数据，接收 long 入参，作为指定长度。

2.4 ByteArrayInputStream 类

ByteArrayInputStream 类包装了一字节数组缓冲区，并提供了字节输入流的相关功能。适用于需要重复使用数据的场景，多线程并发安全。

2.4.1 构造器

ByteArrayInputStream 构造器见表 2-3。

表 2-3 ByteArrayInputStream 构造器

构造器	描述
ByteArrayInputStream(byte[] buf)	构造新的对象，指定缓存数据
ByteArrayInputStream(byte[] buf, int offset, int length)	构造新的对象，指定缓存数据，指定开始索引，指定数据长度

2.4.2 常用方法

本节基于官方文档介绍常用方法，核心方法配合代码示例以方便读者理解。

1. available()

返回从当前输入流中可读取的字节数。

2. close()

关闭当前输入流并释放与此流相关联的任何资源。

3. markSupported()

检查当前输入流是否支持标记和重置方法，返回 boolean 值，代码如下：

```
//第 2 章/four/ByteArrayInputStreamTest.java
public class ByteArrayInputStreamTest {

    public static void main(String[] args) throws IOException {
        byte[] srcByte = new byte[]{65,66,67,68,69,70,71,72,73,74,75,76};
        ByteArrayInputStream byteArrayInputStream = new
                                        ByteArrayInputStream(srcByte);
        System.out.println(byteArrayInputStream.available());
        System.out.println(byteArrayInputStream.markSupported());
    }
}
```

执行结果如下：

```
12
True
```

4. read()

返回从当前输入流中读取的下一字节数据，如果到达流的末尾，则为－1，代码如下：

```
//第 2 章/four/ByteArrayInputStreamTest.java
public class ByteArrayInputStreamTest {

    public static void main(String[] args) throws IOException {
        byte[] srcByte = new byte[]{65,66,67,68,69,70,71,72,73,74,75,76};
        ByteArrayInputStream byteArrayInputStream = new
                                        ByteArrayInputStream(srcByte);
        int bts = byteArrayInputStream.read();
        while (bts != -1){
            System.out.println((char) bts);
            bts = byteArrayInputStream.read();
        }
    }
}
```

执行结果如下：

```
A
B
C
D
E
F
```

```
G
H
I
J
K
L
```

5. read(byte[] b)

从当前输入流中读取字节数据并将它们存储到指定的数组中，返回读取到的字节总数。如果到达此流的末尾，则为－1。接收 byte[]入参，作为指定的数组。

6. readAllBytes()

返回从当前输入流中读取到的所有剩余字节。

7. readNBytes(byte[] b, int off, int len)

从当前输入流中读取字节数据并将它们存储到指定的数组中，返回读取到的字节总数。接收 byte[]入参，作为指定的数组，接收 int 入参，作为数组的起始索引，接收 int 入参，作为要读取的最大字节长度。

8. readNBytes(int len)

返回从当前输入流中读取到的最大长度的字节数据。接收 int 入参，作为要读取的最大字节长度。

9. skip(long n)

跳过当前输入流中指定长度的字节数据，返回已跳过的实际值。接收 long 入参，作为指定长度。

10. mark(int readlimit)

标记当前输入流中数据指针的位置。接收 int 入参，作为在标记位变为无效前可读取的最大字节数。

注意：在当前类中此方法的入参并没有真正地被用到。

---

11. reset()

把当前输入流中的数据指针设置为以前标记的值，代码如下：

```
public class ByteArrayInputStreamTest {

    public static void main(String[] args) throws IOException {
        byte[] srcByte = new byte[]{65,66,67,68,69,70,71,72,73,74,75,76};
        ByteArrayInputStream byteArrayInputStream = new
                                        ByteArrayInputStream(srcByte);
        System.out.println(byteArrayInputStream.available());
```

```
        System.out.println(byteArrayInputStream.markSupported());
        System.out.println(new String(byteArrayInputStream.
                                            readNBytes(3)));
        byteArrayInputStream.mark(454);
        System.out.println(new String(byteArrayInputStream.
                                            readAllBytes()));
        byteArrayInputStream.reset();
        System.out.println(new String(byteArrayInputStream.
                                            readAllBytes()));
        System.out.println(byteArrayInputStream.available());
    }
}
```

执行结果如下：

```
12
true
ABC
DEFGHIJKL
DEFGHIJKL
0
```

11min

2.5 SequenceInputStream 类

SequenceInputStream 类包装了多字节输入流的序列，并提供了字节输入流的相关功能。

2.5.1 构造器

SequenceInputStream 构造器见表 2-4。

表 2-4 SequenceInputStream 构造器

构 造 器	描 述
SequenceInputStream(InputStream s1, InputStream s2)	构造新的对象，指定第 1 字节输入流，指定第 2 字节输入流
SequenceInputStream (Enumeration <? extends InputStream > e)	构造新的对象，指定字节输入流集合

2.5.2 常用方法

本节基于官方文档介绍常用方法，核心方法配合代码示例以方便读者理解。

1. available()

返回从当前输入流中可读取的字节数，代码如下：

```
//第 2 章/five/SequenceInputStreamTest.java
public class SequenceInputStreamTest {

    public static void main(String[] args) {
        try (SequenceInputStream sequenceInputStream =
                                        new SequenceInputStream(
                new FileInputStream("D:\\temp\\src\\test.txt"),
                new FileInputStream("D:\\temp\\src\\test2.txt"))) {
            System.out.println(sequenceInputStream.available());
        } catch (Exception exception) {
            System.out.println(exception);
        }
    }
}
```

执行结果如下：

```
20
```

文件描述信息如图 2-4 所示。

图 2-4 文件描述信息

2. close()

关闭所有输入流并释放与此流相关联的任何资源。

3．markSupported()

检查当前输入流是否支持标记和重置方法，返回 boolean 值，代码如下：

```
//第 2 章/five/SequenceInputStreamTest.java
public class SequenceInputStreamTest {

    public static void main(String[] args) {
        try (SequenceInputStream sequenceInputStream =
                                        new SequenceInputStream(
                new FileInputStream("D:\\temp\\src\\test.txt"),
                new FileInputStream("D:\\temp\\src\\test2.txt"))) {
            System.out.println(sequenceInputStream.markSupported());
        } catch (Exception exception) {
            System.out.println(exception);
        }
    }
}
```

执行结果如下：

```
false
```

4．read()

循环读取包装的多字节输入流，并返回读取的下一字节数据，如果到达最后流的末尾，则为－1，代码如下：

```
//第 2 章/five/SequenceInputStreamTest.java
public class SequenceInputStreamTest {

    public static void main(String[] args) {
        try (SequenceInputStream sequenceInputStream =
                                        new SequenceInputStream(
                new FileInputStream("D:\\temp\\src\\test.txt"),
                new FileInputStream("D:\\temp\\src\\test2.txt"))) {
            int read = sequenceInputStream.read();
            while (read != -1){
                System.out.print(read + " ");
                read = sequenceInputStream.read();
            }
        } catch (Exception exception) {
            System.out.println(exception);
        }
    }
}
```

执行结果如下：

```
65 65 65 229 164 167 231 137 155 232 130 161 66 66 66 67 67 67 13 10 104 101 108 108 111 45 228
184 150 231 149 140 117 117 117
```

5. read(byte[] b)

循环读取包装的多字节输入流,并将读取的字节数据存储到指定的数组中,返回读取到的字节总数。如果到达最后流的末尾,则为-1。接收 byte[]入参,作为指定的数组。

6. readAllBytes()

返回从所有输入流中读取到的所有剩余字节,代码如下:

```
//第 2 章/five/SequenceInputStreamTest.java
public class SequenceInputStreamTest {

    public static void main(String[] args) {
        try (SequenceInputStream sequenceInputStream =
                                                new SequenceInputStream(
                new FileInputStream("D:\\temp\\src\\test.txt"),
                new FileInputStream("D:\\temp\\src\\test2.txt"))) {
            byte[] bytes = sequenceInputStream.readAllBytes();
            System.out.println(new String(bytes));
        } catch (Exception exception) {
            System.out.println(exception);
        }
    }
}
```

执行结果如下:

```
AAA 大牛股 BBBCCC
hello - 世界 uuu
```

文件内容如图 2-5 所示。

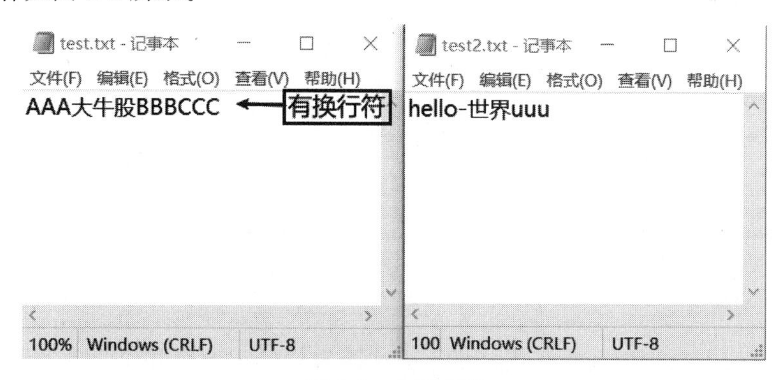

图 2-5 文件内容

7. readNBytes(int len)

返回从所有输入流中读取到的最大长度的字节数据。接收 int 入参,作为要读取的最大字节长度。

2.6 ObjectInputStream 类

ObjectInputStream 类需要配合 ObjectOutputStream 类使用。提供了序列化、反序列化功能,并提供了字节输入流的相关功能。

2.6.1 初始化数据

根据 ObjectOutputStream 类的官方文档示例初始化数据,代码如下:

```
//第 2 章/six/ObjectInputStreamTest.java
public class ObjectInputStreamTest {

    public static void main(String[] args) {
        try (FileOutputStream fos = new
                            FileOutputStream("D:\\temp\\src\\t.txt");
             ObjectOutputStream oos = new ObjectOutputStream(fos)) {
            oos.writeInt(12345);
            oos.writeObject("Today");
            oos.writeObject(new Date());
        } catch (Exception exception) {
            exception.printStackTrace();
        }
    }
}
```

2.6.2 构造器

ObjectInputStream 构造器见表 2-5。

表 2-5 ObjectInputStream 构造器

构 造 器	描 述
ObjectInputStream(InputStream in)	构造新的对象,指定字节输入流

2.6.3 常用方法

本节基于官方文档介绍常用方法,核心方法配合代码示例以方便读者理解。

1. available()

返回从当前输入流中可以不分块读取的字节数,代码如下:

```
//第 2 章/six/ObjectInputStreamTest.java
public class ObjectInputStreamTest {

    public static void main(String[] args) {
```

```
        try (FileInputStream fis = new
                                FileInputStream("D:\\temp\\src\\t.txt");
           ObjectInputStream ois = new ObjectInputStream(fis)) {
          System.out.println(ois.available());
        } catch (Exception e) {
          e.printStackTrace();
        }
    }
}
```

执行结果如下：

```
4
```

2．close()

关闭当前输入流并释放与此流相关联的任何资源。

3．markSupported()

检查当前输入流是否支持标记和重置方法，返回 boolean 值，代码如下：

```
//第 2 章/six/ObjectInputStreamTest.java
public class ObjectInputStreamTest {

    public static void main(String[] args) {
        try (FileInputStream fis = new
                                FileInputStream("D:\\temp\\src\\t.txt");
           ObjectInputStream ois = new ObjectInputStream(fis)) {
          System.out.println(ois.markSupported());
        } catch (Exception e) {
          e.printStackTrace();
        }
    }
}
```

执行结果如下：

```
false
```

4．readBoolean()

返回从当前输入流中读取的下一个 boolean 数据。

5．readChar()

返回从当前输入流中读取的下一个 char 数据。

6．readDouble()

返回从当前输入流中读取的下一个 double 数据。

7．readFloat()

返回从当前输入流中读取的下一个 float 数据。

8. readInt()

返回从当前输入流中读取的下一个 int 数据。

9. readLong()

返回从当前输入流中读取的下一个 long 数据。

10. readShort()

返回从当前输入流中读取的下一个 short 数据。

11. readUTF()

返回从当前输入流中读取的下一个字符串数据。

12. readObject()

返回从当前输入流中读取的下一个 Object 数据，代码如下：

```
//第 2 章/six/ObjectInputStreamTest.java
public class ObjectInputStreamTest {

    public static void main(String[] args) {
        try (FileInputStream fis = new
                                FileInputStream("D:\\temp\\src\\t.txt");
             ObjectInputStream ois = new ObjectInputStream(fis)) {
            int i = ois.readInt();
            System.out.println(i);
            String today = (String) ois.readObject();
            System.out.println(today);
            Date date = (Date) ois.readObject();
            System.out.println(date);
        } catch (Exception e) {
            e.printStackTrace();
        }
    }
}
```

执行结果如下：

```
12345
Today
Wed Nov 08 17:43:36 CST 2023
```

注意：序列化和反序列化的数据存储、读取顺序需要一致。

2.6.4 自定义序列化对象

(1) 序列化对象的类必须实现 Serializable 接口。

(2) 序列化对象的类必须有 serialVersionUID 常量字段。

代码如下：

```
//第 2 章/six/User.java
public class User implements Serializable {
    @java.io.Serial
    @Native
    private static final long serialVersionUID = 1360826667806866666L;
    private String name;
    private String address;
    private Integer age;
    private Integer phone;
//被 transient 关键字修饰的字段不被序列化
    transient private Integer money;

    public User(String name, String address, Integer age,
                                        Integer phone, Integer money) {
        this.name = name;
        this.address = address;
        this.age = age;
        this.phone = phone;
        this.money = money;
    }

    public String getName() {
        return name;
    }

    public void setName(String name) {
        this.name = name;
    }

    public String getAddress() {
        return address;
    }

    public void setAddress(String address) {
        this.address = address;
    }

    public Integer getAge() {
        return age;
    }

    public void setAge(Integer age) {
        this.age = age;
    }

    public Integer getPhone() {
        return phone;
```

```
    }

    public void setPhone(Integer phone) {
        this.phone = phone;
    }

    public Integer getMoney() {
        return money;
    }

    public void setMoney(Integer money) {
        this.money = money;
    }

    @Override
    public String toString() {
        return "User{" +
                "name = '" + name + '\'' +
                ", address = '" + address + '\'' +
                ", age = " + age +
                ", phone = " + phone +
                ", money = " + money +
                '}';
    }
}
```

测试代码如下：

```
//第 2 章/six/ObjectInputStreamTest.java
//import 第 2 章/six/User.java
public class ObjectInputStreamTest {

    public static void main(String[] args) {
        objectWriter();
        objectReader();
    }

    public static void objectReader() {
        try (FileInputStream fis = new
                            FileInputStream("D:\\temp\\src\\t.txt");
             ObjectInputStream ois = new ObjectInputStream(fis)) {
            User user = (User) ois.readObject();
            System.out.println(user);
        } catch (Exception e) {
            e.printStackTrace();
        }
    }

    public static void objectWriter() {
        try (FileOutputStream fos = new
```

```
                FileOutputStream("D:\\temp\\src\\t.txt");
         ObjectOutputStream oos = new ObjectOutputStream(fos)) {
        User user = new User("kungreat", "word", 18, 15454545, 66666);
        oos.writeObject(user);
    } catch (Exception exception) {
        exception.printStackTrace();
    }
  }
}
```

执行结果如下：

```
User{name = 'kungreat', address = 'word', age = 18, phone = 15454545, money = null}
```

在序列化和反序列化过程中需要特殊处理的数据应实现以下方法，如图 2-6 所示。

```
//Date类的源代码
@java.io.Serial
private void writeObject(ObjectOutputStream s)
     throws IOException {
    s.defaultWriteObject();
    s.writeLong(getTimeImpl());
}

@java.io.Serial
private void readObject(ObjectInputStream s)
     throws IOException, ClassNotFoundException {
    s.defaultReadObject();
    fastTime = s.readLong();
}
```

图 2-6　Date 类的源代码

小结

计算机的世界是由二进制数组成的，目前计算机数据存储的最小单位是位(bit)。

习题

1. 判断题

(1) 文件字节输入流可以读取任意类型的文件数据。(　　)

(2) 文件字节输入流使用完毕后必须关闭以释放资源。(　　)

(3) 反序列化必须配合序列化使用。(　　)

(4) 序列化对象可以是任意的对象。(　　)

2. 选择题

（1）InputStream 类可以读取数据的方法有（　　）。（多选）

A. markSupported()　　B. read()

C. readAllBytes()　　D. readNBytes(int len)

（2）InputStream 类可以读取所有剩余数据的方法是（　　）。（单选）

A. markSupported()　　B. read()

C. readAllBytes()　　D. readNBytes(int len)

（3）可以达到字节缓存效果的类是（　　）。（单选）

A. FileInputStream　　B. ObjectInputStream

C. ByteArrayInputStream　　D. SequenceInputStream

（4）包装了多字节输入流序列的类是（　　）。（单选）

A. FileInputStream　　B. ObjectInputStream

C. ByteArrayInputStream　　D. SequenceInputStream

3. 填空题

（1）查看执行结果并补充代码，代码如下：

```
//第 2 章/answer/SerializableTest.java
public class SerializableTest {
   public static void main(String[] args) {
      objectWriter();
      objectReader();
   }

   public static void objectReader() {
      try (FileInputStream fis = new
                        FileInputStream("D:\\temp\\src\\t.txt");
          ObjectInputStream ois = new ObjectInputStream(fis)) {
         int i = ois.________;
         String today = (String) ois.readObject();
         Date date = (Date) ois.readObject();
         System.out.println(i);
         System.out.println(today);
         System.out.println(date);
      } catch (Exception e) {
         e.printStackTrace();
      }
   }

   public static void objectWriter() {
      try (FileOutputStream fos = new
                        FileOutputStream("D:\\temp\\src\\t.txt");
          ObjectOutputStream oos = new ObjectOutputStream(fos)) {
         oos.________________;
```

```
            oos.________________;
            oos.writeObject(new Date(454545454999L));
        } catch (Exception exception) {
            exception.printStackTrace();
        }
    }
}
```

执行结果如下：

```
12345
Today
Mon May 28 06:37:34 CST 1984
```

(2) 查看执行结果并补充代码,代码如下：

```
public class EncodeTest {

    public static void main(String[] args) throws IOException {
        byte[] srcByte = new byte[]{______________________________};
        ByteArrayInputStream byteArrayInputStream = new
                                        ByteArrayInputStream(srcByte);
        System.out.println(byteArrayInputStream.available());
        System.out.println(new
                String(byteArrayInputStream.readNBytes(3)));
        byteArrayInputStream.mark(666);
        System.out.println(new
                        String(byteArrayInputStream.readAllBytes()));
        byteArrayInputStream.reset();
        System.out.println(new
                        String(byteArrayInputStream.readAllBytes()));
        System.out.println(byteArrayInputStream.available());
    }
}
```

执行结果如下：

```
6
ABC
DEF
DEF
0
```

第 3 章

CHAPTER 3

IO 字节输出流

IO 字节输出流通过数据流、序列化、文件系统提供字节输出功能，输入和输出是相对的概念。

4min

3.1 OutputStream 抽象类

此抽象类是所有字节输出流的超类。本节基于官方文档介绍常用方法，核心方法配合代码示例以方便读者理解。

1. close()

关闭当前输出流并释放与此流相关联的任何资源。

2. flush()

清空当前输出流并强制写出任何缓冲的字节。

3. write(byte[] b)

将指定字节数组写出到当前输出流。接收 byte[]入参，作为指定字节数组。

4. write(byte[] b, int off, int len)

将指定字节数组写出到当前输出流。接收 byte[]入参，作为指定字节数组，接收 int 入参，作为数组的起始索引，接收 int 入参，作为写出的最大字节长度。

5. write(int b)

将指定字节数据写出到当前输出流。接收 int 入参，作为指定字节数据。

17min

3.2 FileOutputStream 类

文件字节输出流的作用是将字节数据写出到文件。

3.2.1 构造器

FileOutputStream 构造器见表 3-1。

表 3-1　FileOutputStream 构造器

构　造　器	描　　述
FileOutputStream(File file)	构造新的对象,指定文件系统
FileOutputStream(FileDescriptor fdObj)	构造新的对象,指定文件描述符
FileOutputStream(File file, boolean append)	构造新的对象,指定文件系统,指定是否追加内容
FileOutputStream(String name)	构造新的对象,指定文件系统路径
FileOutputStream(String name, boolean append)	构造新的对象,指定文件系统路径,指定是否追加内容

3.2.2　常用方法

本节基于官方文档介绍常用方法,核心方法配合代码示例以方便读者理解。

1. close()

关闭当前输出流并释放与此流相关联的任何资源。

2. flush()

清空当前输出流并强制写出任何缓冲的字节。

3. write(byte[] b)

将指定字节数组写出到当前输出流。接收 byte[]入参,作为指定字节数组,代码如下:

```
//第 3 章/two/FileOutputStreamTest.java
public class FileOutputStreamTest {

    public static void main(String[] args) {
        try (FileOutputStream fileOutputStream = new
                FileOutputStream("D:\\temp\\src\\test5.txt", false)) {
            String str = "FileOutputStream 文件字节输出流";
            fileOutputStream.write(str.getBytes());
            fileOutputStream.flush();
        } catch (Exception e) {
            e.printStackTrace();
        }
    }
}
```

运行程序后查看文件内容,如图 3-1 所示。

4. write(byte[] b, int off, int len)

将指定字节数组写出到当前输出流。接收 byte[]入参,作为指定字节数组,接收 int 入参,作为数组的起始索引,接收 int 入参,作为写出的最大字节长度。

5. write(int b)

将指定字节数据写出到当前输出流。接收 int 入参,作为指定字节数据。

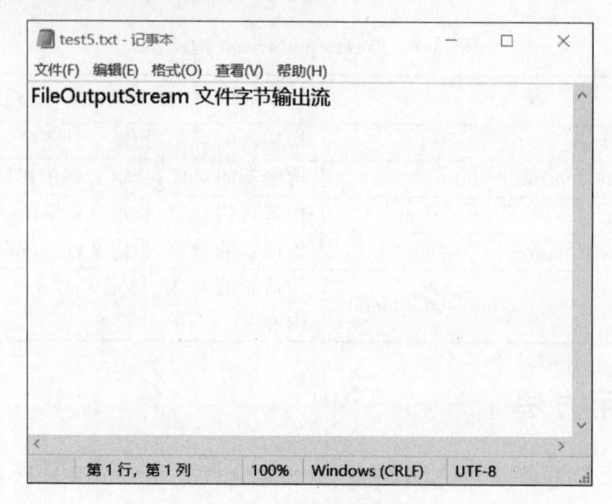

图 3-1　查看文件内容

6．getChannel()

返回与当前文件输出流相关联的唯一文件通道对象。

3.3　ByteArrayOutputStream 类

ByteArrayOutputStream 类包装了一字节数组缓冲区。适用于需要重复使用数据的场景，多线程并发安全。

3.3.1　构造器

ByteArrayOutputStream 构造器见表 3-2。

表 3-2　ByteArrayOutputStream 构造器

构造器	描述
ByteArrayOutputStream()	构造新的对象，默认为无参构造器
ByteArrayOutputStream(int size)	构造新的对象，指定缓冲区的容量大小

3.3.2　常用方法

本节基于官方文档介绍常用方法，核心方法配合代码示例以方便读者理解。

1．close()

关闭当前输出流并释放与此流相关联的任何资源。

2．write(byte[] b)

将指定字节数组写出到当前输出流。接收 byte[]入参，作为指定字节数组，代码如下：

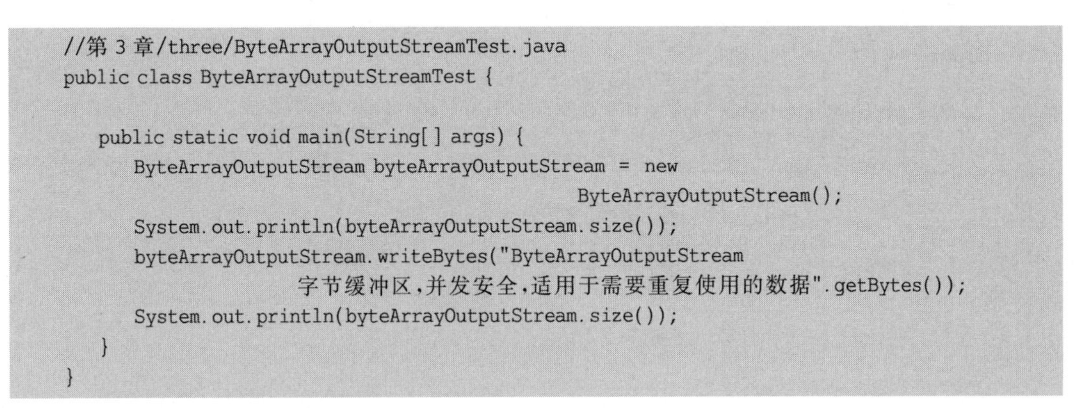

```
//第 3 章/three/ByteArrayOutputStreamTest.java
public class ByteArrayOutputStreamTest {

    public static void main(String[] args) {
        ByteArrayOutputStream byteArrayOutputStream = new
                                            ByteArrayOutputStream();
        System.out.println(byteArrayOutputStream.size());
        byteArrayOutputStream.writeBytes("ByteArrayOutputStream
                    字节缓冲区,并发安全,适用于需要重复使用的数据".getBytes());
        System.out.println(byteArrayOutputStream.size());
    }
}
```

执行结果如下：

```
0
87
```

3. write(byte[] b, int off, int len)

将指定字节数组写出到当前输出流。接收 byte[]入参,作为指定字节数组,接收 int 入参,作为数组的起始索引,接收 int 入参,作为写出的最大字节长度。

4. write(int b)

将指定字节数据写出到当前输出流。接收 int 入参,作为指定字节数据。

5. size()

返回当前缓冲区有效字节的数据长度。

6. reset()

重置当前缓冲区的数据指针,效果类似清空缓冲区的数据。

7. toByteArray()

返回当前缓冲区有效字节数据的副本,代码如下：

```
//第 3 章/three/ByteArrayOutputStreamTest.java
public class ByteArrayOutputStreamTest {
    public static void main(String[] args) {
        ByteArrayOutputStream byteArrayOutputStream = new
                                            ByteArrayOutputStream();
        System.out.println(byteArrayOutputStream.size());
        byteArrayOutputStream.writeBytes("ByteArrayOutputStream
                    字节缓冲区,并发安全,适用于需要重复使用的数据".getBytes());
        System.out.println(byteArrayOutputStream.size());
        try (FileOutputStream fileOutputStream = new
                    FileOutputStream("D:\\temp\\src\\test5.txt", false)) {
            fileOutputStream.write(byteArrayOutputStream.toByteArray());
                //换行符
            fileOutputStream.write(System.lineSeparator().getBytes());
            fileOutputStream.write(byteArrayOutputStream.toByteArray());
```

```
            //换行符
            fileOutputStream.write(System.lineSeparator().getBytes());
            fileOutputStream.write(byteArrayOutputStream.toByteArray());
            fileOutputStream.flush();
        } catch (Exception e) {
            e.printStackTrace();
        }
    }
}
```

运行程序后查看文件内容，如图 3-2 所示。

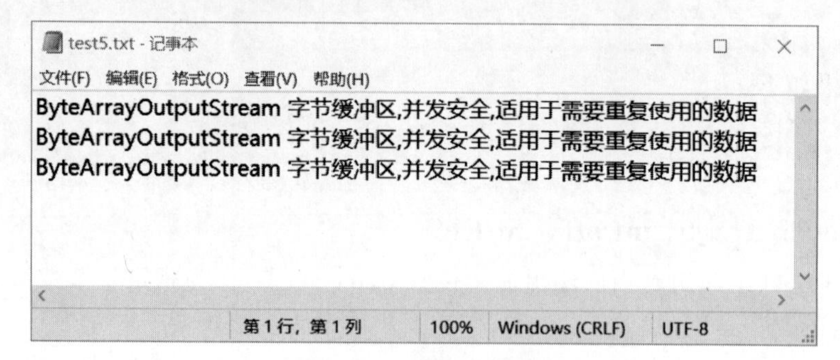

图 3-2　查看文件内容

8. toString()

使用平台的默认字符编码将当前缓冲区内的有效字节数据转换成字符串。

9. toString(Charset charset)

使用指定的字符编码将当前缓冲区内的有效字节数据转换成字符串。接收 Charset 入参，作为指定的字符编码。

10. writeTo(OutputStream out)

将当前缓冲区内的有效字节数据写到指定的输出流。接收 OutputStream 入参，作为指定的输出流。

此方法的源代码如图 3-3 所示。

```
public synchronized void writeTo(OutputStream out) throws IOException {
    out.write(buf, 0, count);
}
```

图 3-3　writeTo(OutputStream out)方法的源代码

3.4　ObjectOutputStream 类

ObjectOutputStream 类需要配合 ObjectInputStream 类使用。提供了序列化、反序列化功能，并提供了字节输出流的相关功能。

3.4.1　构造器

ObjectOutputStream 构造器见表 3-3。

表 3-3　ObjectOutputStream 构造器

构　造　器	描　　述
ObjectOutputStream(OutputStream out)	构造新的对象,指定字节输出流

3.4.2　常用方法

本节基于官方文档介绍常用方法,核心方法配合代码示例以方便读者理解。

1. close()

关闭当前输出流并释放与此流相关联的任何资源。

2. write(byte[] buf)

将指定字节数组写出到当前输出流。接收 byte[]入参,作为指定字节数组。

3. write(byte[] b, int off, int len)

将指定字节数组写出到当前输出流。接收 byte[]入参,作为指定字节数组,接收 int 入参,作为数组的起始索引,接收 int 入参,作为写出的最大字节长度。

4. write(int b)

将指定字节数据写出到当前输出流。接收 int 入参,作为指定字节数据。

5. writeBoolean(boolean val)

将指定布尔数据写出到当前输出流。接收 boolean 入参,作为指定布尔数据。

6. writeChar(int val)

将指定字符数据写出到当前输出流。接收 int 入参,作为指定字符数据。

7. writeDouble(double val)

将指定双精度数据写出到当前输出流。接收 double 入参,作为指定双精度数据。

8. writeFloat(float val)

将指定单精度数据写出到当前输出流。接收 float 入参,作为指定单精度数据。

9. writeInt(int val)

将指定整数型数据写出到当前输出流。接收 int 入参,作为指定整数型数据。

10. writeLong(long val)

将指定长整数型数据写出到当前输出流。接收 long 入参,作为指定长整数型数据。

11. writeObject(Object obj)

将指定对象型数据写出到当前输出流。接收 Object 入参，作为指定对象型数据。

12. writeUTF(String str)

将指定字符串数据写出到当前输出流。接收 String 入参，作为指定字符串数据。

注意：此方法入参的字符串数据默认使用 UTF-8 编码进行字节转换。

28min

3.5 字符编码转换工具类

编写字符文件编码转换工具类，支持任意长度的文件数据，代码如下：

```
//第 3 章/fice/FileEncodeUtil.java
public class FileEncodeUtil {

    public static void encodeChange(File srcFile, String srcEncode,
                                    File targetFile, String targetEncode) {
        if (srcFile == null || srcEncode == null
                                    || targetFile == null || targetEncode == null
                                    || !srcFile.exists() || !targetFile.exists()
                                    || !srcFile.isFile() || !targetFile.isFile()
                                    || !Charset.isSupported(srcEncode)
                                    || !Charset.isSupported(targetEncode)
                                    || srcEncode.equals(targetEncode)) {
            System.out.println("必要入参不支持...");
            return;
        }
        //字节数据缓冲区
        ByteBuffer buffer = ByteBuffer.allocate(8192);
        //源文件解码器
        CharsetDecoder charsetDecoder =
                                    Charset.forName(srcEncode).newDecoder();
        //字符数据缓冲区
        CharBuffer charBuffer = CharBuffer.allocate(8192);
        //解码器的解码结果
        CoderResult coderResult;
        try (FileInputStream fileInputStream = new FileInputStream(srcFile);
            FileOutputStream fileOutputStream = new
                                    FileOutputStream(targetFile, false)) {
            //文件是否还有数据可读
            while (fileInputStream.available() > 0) {
                //文件输入字节流 -> 字节数据缓冲区
                int read = fileInputStream.read(buffer.array(),
                                    buffer.position(), buffer.remaining());
                //设置字节数据缓冲区数据的指针位置
```

```
            buffer.position(read + buffer.position());
            //将字节数据缓冲区切换为输出模式
            buffer.flip();
            do {
                //字节数据缓冲区 -> 解码 -> 字符数据缓冲区
                coderResult = charsetDecoder.decode(buffer,
                                                charBuffer, false);
                if (coderResult.isError()) {
                    throw new UnsupportedEncodingException(
                                                    "解码错误!");
                }
                //将字符数据缓冲区切换为输出模式
                charBuffer.flip();
                //字符数据缓冲区 -> 字符串 -> 编码成字节 -> 目标文件
                fileOutputStream.write(charBuffer.toString().
                        getBytes(Charset.forName(targetEncode)));
                //清空字符数据缓冲区
                charBuffer.clear();
                //判断上次解码时字符数据缓冲区是否容量不够
            //此种情况下字节数据缓冲区可能还有数据可以转换
            } while (coderResult.isOverflow());
            //压缩字节数据缓冲区数据,并切换为输入模式
            buffer.compact();
        }
        if (buffer.position() > 0) {
            throw new UnsupportedEncodingException("解码错误!");
        }
    } catch (Exception exception) {
        exception.printStackTrace();
    }
  }
}
```

注意：以上代码用到了 NIO 相关的知识点，后续章节会详细介绍。

小结

字节输出流可以输出任意格式的文件数据，文件名称后缀是一个规范的设计体系。

习题

1. 判断题

(1) 文件字节输出流可以输出任意类型的文件数据。(　　)

（2）文件字节输出流使用完毕后必须关闭以释放资源。（　　）

（3）字符数据代表文字型内容。（　　）

2. 选择题

（1）OutputStream类可以输出数据的方法有（　　）。（多选）

A. close()　　B. flush()

C. write(byte[] b)　　D. write(int b)

（2）FileOutputStream类可以将数据强制刷新到文件的方法是（　　）。（单选）

A. close()　　B. flush()

C. write(byte[] b)　　D. write(int b)

（3）可以达到字节缓存效果的类是（　　）。（单选）

A. FileOutputStream　　B. ObjectOutputStream

C. ByteArrayOutputStream　　D. PipedOutputStream

（4）可以序列化对象的类是（　　）。（单选）

A. FileOutputStream　　B. ObjectOutputStream

C. ByteArrayOutputStream　　D. PipedOutputStream

3. 填空题

查看执行结果并补充代码，代码如下：

```
//第3章/answer/SerializableTest.java
public class SerializableTest {
    public static void main(String[] args) {
        objectWriter();
        objectReader();
    }

    public static void objectReader() {
        try (FileInputStream fis = new
                            FileInputStream("D:\\temp\\src\\t.txt");
             ObjectInputStream ois = new ObjectInputStream(fis)) {
            int i = ois.readInt();
            String today = (String) ois.________;
            Date date = (Date) ois.________;
            System.out.println(i);
            System.out.println(today);
            System.out.println(date);
        } catch (Exception e) {
            e.printStackTrace();
        }
    }

    public static void objectWriter() {
        try (FileOutputStream fos = new
                            FileOutputStream("D:\\temp\\src\\t.txt");
```

```
            ObjectOutputStream oos = new ObjectOutputStream(fos)) {
            oos.writeInt(12345);
            oos.________________
            oos.writeObject(new Date(454545454999L));
        } catch (Exception exception) {
            exception.printStackTrace();
        }
    }
}
```

执行结果如下：

```
12345
Today
Mon May 28 06:37:34 CST 1984
```

第4章 IO字符输入流

CHAPTER 4

IO字符输入流通过数据流、文件系统提供字符输入功能。

4.1 Reader抽象类

此抽象类定义了把字节数据转换为字符数据的规范方法。本节基于官方文档介绍常用方法，核心方法配合代码示例以方便读者理解。

1. close()

关闭当前输入流并释放与此流相关联的任何资源。

2. mark(int readAheadLimit)

标记当前输入流中数据指针的位置。接收int入参，作为在标记位变为无效前可读取的最大字符数。

3. markSupported()

检查当前输入流是否支持标记和重置方法，返回boolean值。

4. read()

返回从当前输入流中读取的下一个字符数据。

5. read(char[]cbuf)

从当前输入流中读取字符数据并将它们存储到指定的数组中，返回读取到的字符总数。如果到达此流的末尾，则为-1。接收char[]入参，作为指定的数组。

6. read(char[] cbuf, int off, int len)

从当前输入流中读取字符数据并将它们存储到指定的数组中，返回读取到的字符总数。接收char[]入参，作为指定的数组，接收int入参，作为数组的起始索引，接收int入参，作为要读取的最大字符长度。

7. read(CharBuffer target)

从当前输入流中读取字符数据并将它们存储到指定的字符缓冲区，返回读取到的字符总数。如果到达此流的末尾,则为-1。接收 CharBuffer 入参,作为指定的字符缓冲区。

8. ready()

检查当前输入流是否已准备就绪,返回 boolean 值。

9. reset()

把当前输入流中的数据指针设置为以前标记的值。

10. skip(long n)

跳过当前输入流中指定长度的字符数据,返回已跳过的实际值。接收 long 入参,作为指定长度。

4.2 InputStreamReader 类

15min

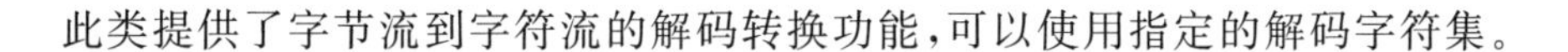

此类提供了字节流到字符流的解码转换功能,可以使用指定的解码字符集。

4.2.1 构造器

InputStreamReader 构造器见表 4-1。

表 4-1 InputStreamReader 构造器

构造器	描述
InputStreamReader(InputStream in)	构造新的对象,指定字节输入流
InputStreamReader(InputStream in, String charsetName)	构造新的对象,指定字节输入流,指定字符集名称
InputStreamReader(InputStream in, Charset cs)	构造新的对象,指定字节输入流,指定字符集
InputStreamReader(InputStream in, CharsetDecoder dec)	构造新的对象,指定字节输入流,指定字符集解码器

4.2.2 常用方法

本节基于官方文档介绍常用方法,核心方法配合代码示例以方便读者理解。

1. close()

关闭当前输入流并释放与此流相关联的任何资源。

2. mark(int readAheadLimit)

标记当前输入流中数据指针的位置。接收 int 入参,作为在标记位变为无效前可读取的最大字符数。

3. markSupported()

检查当前输入流是否支持标记和重置方法,返回 boolean 值,代码如下:

```
//第 4 章/two/InputStreamReaderTest.java
public class InputStreamReaderTest {

    public static void main(String[] args) {
        try (InputStreamReader reader = new InputStreamReader
                (new FileInputStream("D:\\temp\\src\\test.txt"),
                                              StandardCharsets.UTF_8)) {
            System.out.println(reader.getEncoding());
            System.out.println(reader.markSupported());
        } catch (Exception e) {
            e.printStackTrace();
        }

    }
}
```

执行结果如下：

```
UTF8
false
```

4. read()

返回从当前输入流中读取的下一个字符数据，代码如下：

```
//第 4 章/two/InputStreamReaderTest.java
public class InputStreamReaderTest {

    public static void main(String[] args) {
        try (InputStreamReader inputStreamReader = new InputStreamReader(
                new FileInputStream("D:\\temp\\src\\test.txt"),
                                      StandardCharsets.UTF_8)) {
            int read = inputStreamReader.read();
            while (read != -1){
                System.out.println((char)read);
                read = inputStreamReader.read();
            }
        } catch (Exception e) {
            e.printStackTrace();
        }

    }
}
```

执行结果如下：

```
A
A
A
大
牛
```

```
股
B
B
B
C
C
C
```

文件内容如图 4-1 所示。

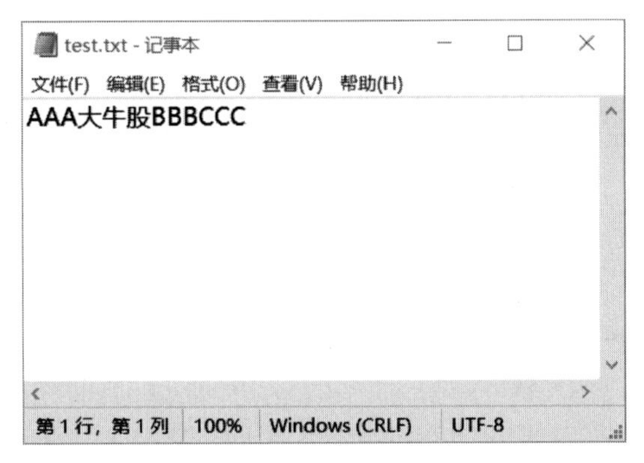

图 4-1 文件内容

5. read(char[] cbuf)

从当前输入流中读取字符数据并将它们存储到指定的数组中，返回读取到的字符总数。如果到达此流的末尾，则为－1。接收 char[]入参，作为指定的数组，代码如下：

```java
//第 4 章/two/InputStreamReaderTest.java
public class InputStreamReaderTest {

    public static void main(String[] args) {
        try (InputStreamReader inputStreamReader = new InputStreamReader(
                new FileInputStream("D:\\temp\\src\\test.txt"),
                                                StandardCharsets.UTF_8)) {
            char[] chars = new char[666];
            int read = inputStreamReader.read(chars);
            System.out.println(new String(chars, 0, read));
        } catch (Exception e) {
            e.printStackTrace();
        }

    }
}
```

执行结果如下：

```
AAA 大牛股 BBBCCC
```

6. read(char[] cbuf, int off, int len)

从当前输入流中读取字符数据并将它们存储到指定的数组中，返回读取到的字符总数。接收 char[]入参，作为指定的数组，接收 int 入参，作为数组的起始索引，接收 int 入参，作为要读取的最大字符长度。

7. read(CharBuffer target)

从当前输入流中读取字符数据并将它们存储到指定的字符缓冲区，返回读取到的字符总数。如果到达此流的末尾，则为-1。接收 CharBuffer 入参，作为指定的字符缓冲区。

8. ready()

检查当前输入流是否已准备就绪，返回 boolean 值。

9. reset()

把当前输入流中的数据指针设置为以前标记的值。

10. skip(long n)

跳过当前输入流中指定长度的字符数据，返回已跳过的实际值。接收 long 入参，作为指定长度，代码如下：

```
//第 4 章/two/InputStreamReaderTest.java
public class InputStreamReaderTest {

    public static void main(String[] args) {
        try (InputStreamReader inputStreamReader = new InputStreamReader(
                new FileInputStream("D:\\temp\\src\\test.txt"),
                                                StandardCharsets.UTF_8)) {
            System.out.println(inputStreamReader.skip(4));
            char[] chars = new char[666];
            int read = inputStreamReader.read(chars);
            System.out.println(new String(chars, 0, read));
        } catch (Exception e) {
            e.printStackTrace();
        }

    }
}
```

执行结果如下：

```
4
牛股 BBBCCC
```

12min

4.3 BufferedReader 类

BufferedReader 类包装了一个字符数组缓冲区，并提供了字符输入流的相关功能。多线程并发安全。

4.3.1 构造器

BufferedReader 构造器见表 4-2。

表 4-2 BufferedReader 构造器

构造器	描述
BufferedReader(Reader in)	构造新的对象,指定字符输入流
BufferedReader(Reader in, int sz)	构造新的对象,指定字符输入流,指定缓冲区的容量大小

4.3.2 常用方法

本节基于官方文档介绍常用方法,核心方法配合代码示例以方便读者理解。

1. close()

关闭当前输入流并释放与此流相关联的任何资源。

2. mark(int readAheadLimit)

标记当前输入流中数据指针的位置。接收 int 入参,作为在标记位变为无效前可读取的最大字符数。

3. markSupported()

检查当前输入流是否支持标记和重置方法,返回 boolean 值。

4. read()

返回从当前输入流中读取的下一个字符数据。

5. read(char[] cbuf)

从当前输入流中读取字符数据并将它们存储到指定的数组中,返回读取到的字符总数。如果到达此流的末尾,则为-1。接收 char[]入参,作为指定的数组。

6. read(char[] cbuf, int off, int len)

从当前输入流中读取字符数据并将它们存储到指定的数组中,返回读取到的字符总数。接收 char[]入参,作为指定的数组,接收 int 入参,作为数组的起始索引,接收 int 入参,作为要读取的最大字符长度。

7. ready()

检查当前输入流是否已准备就绪,返回 boolean 值。

8. reset()

把当前输入流中的数据指针设置为以前标记的值。

9. skip(long n)

跳过当前输入流中指定长度的字符数据，返回已跳过的实际值。接收 long 入参，作为指定长度。

10. readLine()

返回从当前输入流中读取的下一行字符串数据，不包含行终止符。如果到达流的末尾而未读取到任何字符，则为空，代码如下：

```
//第 4 章/three/BufferedReaderTest.java
public class BufferedReaderTest {

    public static void main(String[] args) {
        try (InputStreamReader inputStreamReader = new InputStreamReader(
                new FileInputStream("D:\\temp\\src\\test5.txt"), "UTF-8");
             BufferedReader bufferedReader = new
                                    BufferedReader(inputStreamReader)) {
            System.out.println(bufferedReader.markSupported());
            System.out.println(bufferedReader.readLine());
            System.out.println(bufferedReader.readLine());
            System.out.println(bufferedReader.readLine());
            System.out.println(bufferedReader.readLine());
        } catch (Exception e) {
            e.printStackTrace();
        }
    }
}
```

执行结果如下：

```
true
ByteArrayOutputStream 字节缓冲区,并发安全,适用于需要重复使用的数据
ByteArrayOutputStream 字节缓冲区,并发安全,适用于需要重复使用的数据
ByteArrayOutputStream 字节缓冲区,并发安全,适用于需要重复使用的数据
null
```

文件内容如图 4-2 所示。

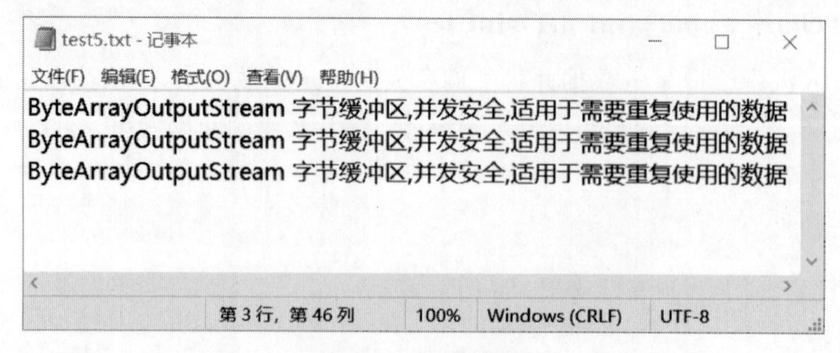

图 4-2 文件内容

11. lines()

返回从当前输入流中读取的所有字符串数据,以行终止符分割,代码如下:

```
//第 4 章/three/BufferedReaderTest.java
public class BufferedReaderTest {

    public static void main(String[] args) {
        try (InputStreamReader inputStreamReader = new InputStreamReader(
                new FileInputStream("D:\\temp\\src\\test5.txt"), "UTF-8");
             BufferedReader bufferedReader = new
                                 BufferedReader(inputStreamReader)) {
            Stream<String> lines = bufferedReader.lines();
            lines.forEach(System.out::println);
        } catch (Exception e) {
            e.printStackTrace();
        }
    }
}
```

执行结果如下:

```
ByteArrayOutputStream 字节缓冲区,并发安全,适用于需要重复使用的数据
ByteArrayOutputStream 字节缓冲区,并发安全,适用于需要重复使用的数据
ByteArrayOutputStream 字节缓冲区,并发安全,适用于需要重复使用的数据
```

小结

字符输入流用于读取文字型数据,并封装了字节数据转换为字符数据的功能。

习题

1. 判断题

(1) 字符输入流可以读取图片型文件数据。(　　)

(2) 字符代表文字型数据。(　　)

(3) 字节代表任意型数据。(　　)

2. 选择题

(1) Reader 抽象类的抽象方法有(　　)。(多选)

A. close()　　B. read()

C. read(char[] cbuf, int off, int len)　　D. reset()

(2) 可以达到字符缓存效果的类是(　　)。(单选)

A. InputStreamReader　　B. BufferedReader

C. FileReader　　D. PipedReader

(3) 可以达到字节缓存效果的类有(　　)。(多选)

A. InputStreamReader　　B. BufferedReader

C. FileReader　　D. PipedReader

3. 填空题

根据业务要求补全代码，读取文件所有数据并以行终止符分割，代码如下：

```
public class OneAnswer {
    public static void main(String[] args) {
        try (InputStreamReader inputStreamReader = new InputStreamReader(
                new FileInputStream("D:\\temp\\src\\test5.txt"));
             BufferedReader bufferedReader = new
                                    BufferedReader(inputStreamReader)) {
            Stream<String> lines = ____________________
            lines.forEach(System.out::println);
        } catch (Exception e) {
            e.printStackTrace();
        }
    }
}
```

第5章 CHAPTER 5 IO字符输出流

IO 字符输出流通过数据流、文件系统提供字符输出功能。

5.1 Writer 抽象类

此抽象类定义了把字符数据转换为字节数据的规范方法。本节基于官方文档介绍常用方法,核心方法配合代码示例以方便读者理解。

1. append(char c)

将指定的字符添加到当前字符输出流中。接收 char 入参,作为指定的字符。

2. append(CharSequence csq)

将指定的字符序列添加到当前字符输出流中。接收 CharSequence 入参,作为指定的字符序列。

3. close()

关闭当前输出流并释放与此流相关联的任何资源。

4. flush()

清空当前输出流并强制写出任何缓冲的字符。

5. write(char[] cbuf)

将指定字符数组写出到当前输出流。接收 char[]入参,作为指定字符数组。

6. write(char[] cbuf, int off, int len)

将指定字符数组写出到当前输出流。接收 char[]入参,作为指定字符数组,接收 int 入参,作为数组的起始索引,接收 int 入参,作为写出的最大字符长度。

7. write(int c)

将指定字符数据写出到当前输出流。接收 int 入参,作为指定字符数据。

8. write(String str)

将指定字符串写出到当前输出流。接收 String 入参,作为指定字符串。

16min

5.2 OutputStreamWriter 类

此类提供了字符流到字节流的编解码转换功能,可以使用指定的编解码字符集。

5.2.1 构造器

OutputStreamWriter 构造器见表 5-1。

表 5-1 OutputStreamWriter 构造器

构 造 器	描 述
OutputStreamWriter(OutputStream out)	构造新的对象,指定字节输出流
OutputStreamWriter(OutputStream out, String charsetName)	构造新的对象,指定字节输出流,指定字符集名称
OutputStreamWriter(OutputStream out, Charset cs)	构造新的对象,指定字节输出流,指定字符集
OutputStreamWriter(OutputStream out, CharsetEncoder enc)	构造新的对象,指定字节输出流,指定字符集编码器

5.2.2 常用方法

本节基于官方文档介绍常用方法,核心方法配合代码示例以方便读者理解。

1. close()

关闭当前输出流并释放与此流相关联的任何资源。

2. flush()

清空当前输出流并强制写出任何缓冲的字符。

3. getEncoding()

返回当前输出流所使用的字符集编码的名称。

4. append(char c)

将指定的字符添加到当前字符输出流中。接收 char 入参,作为指定的字符。

5. append(CharSequence csq)

将指定的字符序列添加到当前字符输出流中。接收 CharSequence 入参,作为指定的字符序列,代码如下:

```
//第 5 章/two/OutputStreamWriterTest.java
public class OutputStreamWriterTest {

    public static void main(String[] args) {
```

```
        try (OutputStreamWriter out = new OutputStreamWriter(
                new FileOutputStream("D:\\temp\\src\\test.txt"))) {
            out.write("hello world");
            out.append(System.lineSeparator()).append("end");
            out.flush();
        }catch (Exception e){
            e.printStackTrace();
        }
    }
}
```

运行程序后查看文件内容,如图 5-1 所示。

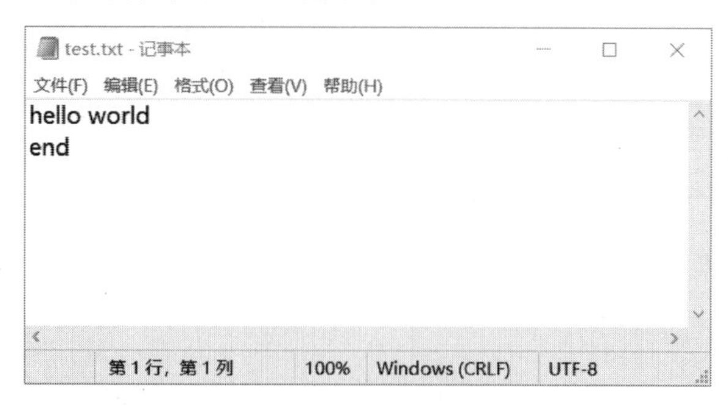

图 5-1　文件内容

6. write(char[] cbuf)

将指定字符数组写出到当前输出流。接收 char[]入参,作为指定字符数组。

7. write(char[] cbuf, int off, int len)

将指定字符数组写出到当前输出流。接收 char[]入参,作为指定字符数组,接收 int 入参,作为数组的起始索引,接收 int 入参,作为写出的最大字符长度。

8. write(int c)

将指定字符数据写出到当前输出流。接收 int 入参,作为指定字符数据,代码如下:

```
//第 5 章/two/OutputStreamWriterTest.java
public class OutputStreamWriterTest {

    public static void main(String[] args) {
        try (OutputStreamWriter out = new OutputStreamWriter(
                new FileOutputStream("D:\\temp\\src\\test.txt"))) {
            out.write(55401);
            out.write(57112);
//"\uD869\uDF18"
        }catch (Exception e){
            e.printStackTrace();
        }
    }
}
```

运行程序后查看文件内容，如图5-2所示。

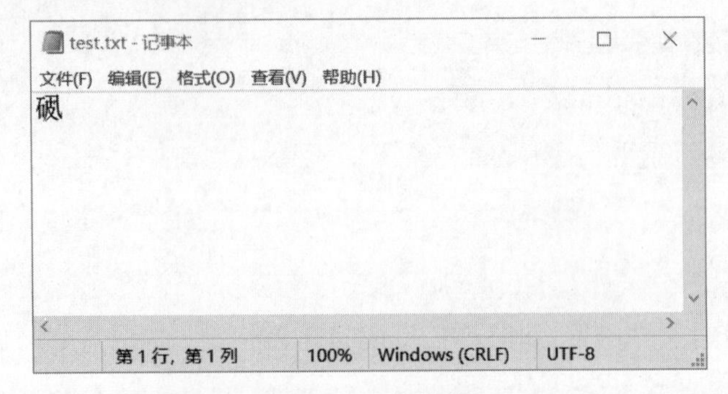

图5-2 文件内容

注意：此文字属于是Unicode扩展区文字。

9. write(String str)

将指定字符串写出到当前输出流。接收String入参，作为指定字符串。

5.3 CharArrayWriter类

CharArrayWriter类包装了一个字符数组缓冲区。适用于需要重复使用数据的场景，多线程并发安全。

5.3.1 构造器

CharArrayWriter构造器见表5-2。

表5-2 CharArrayWriter构造器

构 造 器	描 述
CharArrayWriter()	构造新的对象，默认为无参构造器
CharArrayWriter(int initialSize)	构造新的对象，指定缓冲区的容量大小

5.3.2 常用方法

本节基于官方文档介绍常用方法，核心方法配合代码示例以方便读者理解。

1. close()

此方法空实现。

2. flush()

此方法空实现。

3. append(char c)

将指定的字符添加到当前字符输出流中。接收 char 入参,作为指定的字符。

4. append(CharSequence csq)

将指定的字符序列添加到当前字符输出流中。接收 CharSequence 入参,作为指定的字符序列。

5. write(char[] cbuf)

将指定字符数组写出到当前输出流。接收 char[]入参,作为指定字符数组。

6. write(char[] cbuf, int off, int len)

将指定字符数组写出到当前输出流。接收 char[]入参,作为指定字符数组,接收 int 入参,作为数组的起始索引,接收 int 入参,作为写出的最大字符长度。

7. write(int c)

将指定字符数据写出到当前输出流。接收 int 入参,作为指定字符数据。

8. write(String str)

将指定字符串写出到当前输出流。接收 String 入参,作为指定字符串。

9. writeTo(Writer out)

将当前缓冲区内的有效字符数据写到指定的输出流。接收 Writer 入参,作为指定的输出流,代码如下:

```
//第 5 章/three/CharArrayWriterTest.java
public class CharArrayWriterTest {

    public static void main(String[] args) throws IOException {
        CharArrayWriter charArrayWriter = new CharArrayWriter();
        charArrayWriter.append("hello").append("world");
        charArrayWriter.write(System.lineSeparator());
        charArrayWriter.write("end");
        System.out.println(charArrayWriter);

        try (OutputStreamWriter out = new OutputStreamWriter(
                new FileOutputStream("D:\\temp\\src\\test.txt"))) {
            charArrayWriter.writeTo(out);
            charArrayWriter.writeTo(out);
            charArrayWriter.writeTo(out);
        }catch (Exception e){
            e.printStackTrace();
        }
    }
}
```

执行结果如下:

```
helloworld
end
```

运行程序后查看文件内容，如图 5-3 所示。

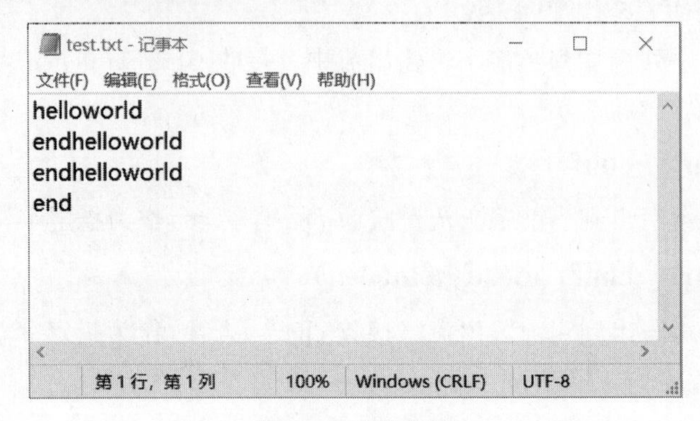

图 5-3 文件内容

10. size()

返回当前缓冲区有效字符的数据长度。

11. reset()

重置当前缓冲区的数据指针，效果类似清空缓冲区的数据。

小结

字符输出流用于输出文字型数据，并封装了字符数据转换为字节数据的功能。

习题

1. 判断题

(1) 单个 char 型数据可能无法表示一个完整的文字。(　　)

(2) 字符编码和解码必须使用相同的字符集。(　　)

(3) 使用 UTF-8 编码的数据可以用 GBK 解码。(　　)

2. 选择题

(1) Writer 抽象类的抽象方法有(　　)。(多选)

A. close()　　B. flush()

C. write(char[] cbuf, int off, int len)　　D. write(String str)

(2) 可以达到字符缓存效果的类是(　　)。(单选)

A. OutputStreamWriter　　B. CharArrayWriter

C. FileWriter D. Object

(3) 常用的字符集编码有（ ）。（多选）

A. UTF-8 B. GBK

C. UTF-16 D. Unicode

3. 填空题

查看执行结果并补充代码,代码如下：

```
//第 5 章/answer/OneAnswer.java
public class OneAnswer {
    public static void main(String[] args) throws IOException {
        CharArrayWriter charArrayWriter = new CharArrayWriter();
        charArrayWriter.append("hello").append("world");
        charArrayWriter.write(System.________________);
        charArrayWriter.write("end");
        System.out.println(charArrayWriter);
        charArrayWriter.________________;
        System.out.println(charArrayWriter);
    }
}
```

执行结果如下：

```
helloworld
end
```

第6章 反射机制

CHAPTER 6

通过反射机制可以解决很多硬编码解决不了的问题，反射机制被称为框架基石。

52min

6.1 Class 类

Class 类封装了底层类的描述信息，包括枚举类、记录类、注解、接口、数组、基本类型。

6.1.1 使用场景

如果要设计一个通用的框架供第三方使用，第三方的使用方式则会存在很多未知因素。通过硬编码的形式是不能解决这种问题的，但可以使用反射机制来处理这种问题，代码如下：

```
//第 6 章/one/Reflex.java
public class Reflex {

    public static void main(String[] args) throws Exception {
        runEverything(new Reflex(), "tsObject", "run - Object");
        runEverything(new Reflex(), "tsStatic", "run - static");
    }

    public void tsObject(String st) {
        System.out.println("ts::" + st);
    }

    public static void tsStatic(String st) {
        System.out.println("ts::" + st);
    }
    //调用任意对象的指定名称的方法(带方法入参)
    public static void runEverything(Object obj, String methodName,
                                     Object... args) throws Exception{
        Class<?> aClass = obj.getClass();
        if (!aClass.isPrimitive()) {
            Method[] declaredMethods = aClass.getDeclaredMethods();
```

```
            for (Method methods : declaredMethods) {
                String name = methods.getName();
                if (name.equals(methodName)) {
                    if (Modifier.isStatic(methods.getModifiers())) {
                        System.out.println("isStatic = true");
                        methods.invoke(null, args);
                    } else {
                        System.out.println("isStatic = false");
                        methods.invoke(obj, args);
                    }
                }
            }
        }
    }
}
```

执行结果如下：

```
isStatic = false
ts::run - Object
isStatic = true
ts::run - static
```

6.1.2 类型信息

通过 Class < T >类可以获取底层包装类的类型信息。

1. 基本类型

基本类型演示，代码如下：

```
//第 6 章/one/TypeTest.java
public class TypeTest {

    public static void main(String[] args) {
        Class <?> intCls = int.class;
        Class <?> voidCls = void.class;
        Class <?> doubleCls = double.class;
        Class <?> longCls = long.class;
        Class <?> shortCls = short.class;
        Class <?> byteCls = byte.class;
        Class <?> floatCls = float.class;
        Class <?> booleanCls = boolean.class;
        Class <?> charCls = char.class;
        System.out.println(intCls);
        System.out.println(voidCls);
        System.out.println(doubleCls);
        System.out.println(voidCls.isPrimitive());
    }

}
```

执行结果如下：

```
int
void
double
true
```

2. 对象类型

对象类型演示，代码如下：

```java
//第6章/one/TypeTest.java
public class TypeTest {

    public static void main(String[] args) {
        Class<?> intCls1 = Integer.class;
        Class<?> intCls2 = Integer.valueOf(55).getClass();
        System.out.println(intCls1);
        System.out.println(intCls1 == intCls2);
        System.out.println(intCls1.isPrimitive());
    }

}
```

执行结果如下：

```
class java.lang.Integer
true
false
```

3. 枚举类型

枚举类型演示，代码如下：

```java
//第6章/one/TypeTest.java
public class TypeTest {

    public static void main(String[] args) {
        Class<?> enumTest1 = EnumTest.class;
        Class<?> enumTest2 = EnumTest.ONE.getClass();
        System.out.println(enumTest1);
        System.out.println(enumTest1 == enumTest2);
        //是否是枚举类
        System.out.println(enumTest1.isEnum());
    }

    private enum EnumTest{
        ONE,TWO;
    }
}
```

执行结果如下：

```
class cn.kungreat.niobook.six.one.TypeTest$EnumTest
true
true
```

4. 接口类型

接口类型演示,代码如下:

```
//第 6 章/one/TypeTest.java
public class TypeTest {

    public static void main(String[] args) {
        Class<?> listClass = List.class;
        System.out.println(listClass);
        //是否是接口类型
        System.out.println(listClass.isInterface());
    }
}
```

执行结果如下:

```
interface java.util.List
true
```

5. 记录类型

记录类型演示,代码如下:

```
//第 6 章/one/TypeTest.java
public class TypeTest {

    public static void main(String[] args) {
        Class<?> recodeTestClass = RecodeTest.class;
        System.out.println(recodeTestClass);
        //是否是记录类型
        System.out.println(recodeTestClass.isRecord());
        //是否是内部类
        System.out.println(recodeTestClass.isMemberClass());
    }

    private record RecodeTest(String name, Integer age){}
}
```

执行结果如下:

```
class cn.kungreat.niobook.six.one.TypeTest$RecodeTest
true
true
```

6. 数组类型

数组类型演示,代码如下:

```java
//第 6 章/one/TypeTest.java
public class TypeTest {

    public static void main(String[] args) {
        Class<?> intCls = int[].class;
        System.out.println(intCls);
        //是否是数组类型
        System.out.println(intCls.isArray());
        if(intCls.isArray()){
            //底层包装的类型
            System.out.println(intCls.componentType());
        }
        //是否是内部类
        System.out.println(intCls.isMemberClass());
    }

}
```

执行结果如下：

```
class [I
true
int
false
```

7. 注解类型

注解类型演示，代码如下：

```java
//第 6 章/one/TypeTest.java
public class TypeTest {

    public static void main(String[] args) {
        Class<?> overrideClass = Override.class;
        System.out.println(overrideClass);
        //是否是注解类型
        System.out.println(overrideClass.isAnnotation());
        //是否是内部类
        System.out.println(overrideClass.isMemberClass());
    }

}
```

执行结果如下：

```
interface java.lang.Override
true
false
```

8. 匿名内部类类型

匿名内部类类型演示，代码如下：

```java
//第 6 章/one/TypeTest.java
public class TypeTest {

    public static void main(String[] args) {
        Class<?> anonymousClass = new Runnable() {
            @Override
            public void run() {
            }
        }.getClass();
        //是否是匿名内部类
        System.out.println(anonymousClass.isAnonymousClass());
        System.out.println(anonymousClass.isMemberClass());
    }

}
```

执行结果如下：

```
true
false
```

6.1.3 元数据信息

通过 Class< T >类可以获取底层包装类的元数据信息，代码如下：

```java
//第 6 章/one/MetaDataTest.java
public class MetaDataTest {

    public static void main(String[] args) {
        Class<MetaDataInner> metaDataInnerClass = MetaDataInner.class;
        //1.继承的超类
        System.out.println(metaDataInnerClass.getGenericSuperclass());
        System.out.println(metaDataInnerClass.getSuperclass());
        //2.实现的接口
        System.out.println(Arrays.toString(metaDataInnerClass.
                                        getInterfaces()));
        //3.类上的注解
        System.out.println(Arrays.toString(metaDataInnerClass.
                                getDeclaredAnnotations()));
        //4.构造器
        System.out.println(Arrays.toString(metaDataInnerClass.
                                        getDeclaredConstructors()));
        //5.内部类
        System.out.println(Arrays.toString(MetaDataTest.class.
                                        getDeclaredClasses()));
        //6.所有方法
        System.out.println(Arrays.toString(metaDataInnerClass.
                                                getDeclaredMethods()));
        //7.所有字段
```

```
        System.out.println(Arrays.toString(metaDataInnerClass.
                                                    getDeclaredFields()));
        //8.包名称
        System.out.println(metaDataInnerClass.getPackageName());
        //简单类名称
        System.out.println(metaDataInnerClass.getSimpleName());
    }

    @AnnotationTest({"ffgdf","dfdsfd"})
    private static final class MetaDataInner implements
                                                    Comparator<MetaDataInner> {
        private String name;

        private Integer age;

        public MetaDataInner(String name, Integer age) {
            this.name = name;
            this.age = age;
        }

        public String getName() {
            return name;
        }

        public void setName(String name) {
            this.name = name;
        }

        public Integer getAge() {
            return age;
        }

        public void setAge(Integer age) {
            this.age = age;
        }

        @Override
        public int compare(MetaDataInner o1, MetaDataInner o2) {
            return 0;
        }
    }

}
```

执行结果如下：

```
class java.lang.Object
class java.lang.Object
[interface java.util.Comparator]
[@cn.kungreat.niobook.six.one.AnnotationTest({"ffgdf", "dfdsfd"})]
```

```
[public cn.kungreat.niobook.six.one.MetaDataTest$MetaDataInner
                                (java.lang.String,java.lang.Integer)]
[class cn.kungreat.niobook.six.one.MetaDataTest$MetaDataInner]
[public java.lang.String
cn.kungreat.niobook.six.one.MetaDataTest$MetaDataInner.getName(),
public int
cn.kungreat.niobook.six.one.MetaDataTest$MetaDataInner.compare(java.lang.Object,java.
lang.Object), public int
cn.kungreat.niobook.six.one.MetaDataTest$MetaDataInner.compare(cn.kungreat.niobook.six.
one.MetaDataTest$MetaDataInner,
cn.kungreat.niobook.six.one.MetaDataTest$MetaDataInner), public void
cn.kungreat.niobook.six.one.MetaDataTest$MetaDataInner.setName(java.lang.String), public
java.lang.Integer
cn.kungreat.niobook.six.one.MetaDataTest$MetaDataInner.getAge(), public void cn.kungreat.
niobook.six.one.MetaDataTest$MetaDataInner.setAge(java.lang.Integer)]
[private java.lang.String
cn.kungreat.niobook.six.one.MetaDataTest$MetaDataInner.name, private java.lang.Integer cn.
kungreat.niobook.six.one.MetaDataTest$MetaDataInner.age]
cn.kungreat.niobook.six.one
MetaDataInner
```

6.2 Constructor 类

Constructor类提供了类单个构造函数的相关信息和访问权。本节基于官方文档介绍常用方法,核心方法配合代码示例以方便读者理解。

1. getAnnotatedReturnType()

返回此构造函数的对象类型。

2. getDeclaredAnnotations()

返回此构造函数的注解信息,代码如下:

```java
//第 6 章/two/ConstructorTest.java
public class ConstructorTest {
    public static void main(String[] args) throws Exception {
        Constructor<?>[] constructors =
                MetaDataInner.class.getDeclaredConstructors();
        System.out.println(Arrays.toString(constructors));
        Constructor<?> constructor = constructors[0];
        System.out.println(Arrays.toString(
                        constructor.getDeclaredAnnotations()));
    }

    public static final class MetaDataInner {
        private String name;

        private Integer age;

        @AnnotationTest({"construct"})
        public MetaDataInner(@AnnotationTest({"param1"}) String name,
```

```
            @AnnotationTest({"param2"}) Integer age) throws Exception{
            this.name = name;
            this.age = age;
        }

        public String getName() {
            return name;
        }

        public void setName(String name) {
            this.name = name;
        }

        public Integer getAge() {
            return age;
        }

        public void setAge(Integer age) {
            this.age = age;
        }
    }
}
```

执行结果如下：

```
[public cn.kungreat.niobook.six.two.ConstructorTest$MetaDataInner(
    java.lang.String,java.lang.Integer) throws java.lang.Exception]
[@cn.kungreat.niobook.six.one.AnnotationTest({"construct"})]
```

3. getDeclaringClass()

返回此构造函数类的描述对象。

4. getExceptionTypes()

返回此构造函数所抛出的异常类型。

5. getParameterTypes()

返回此构造函数的入参信息，代码如下：

```
/第6章/two/ConstructorTest.java
public class ConstructorTest {
    public static void main(String[] args) throws Exception {
        Constructor<?>[] constructors =
                MetaDataInner.class.getDeclaredConstructors();
        Constructor<?> constructor = constructors[0];
        System.out.println(Arrays.toString(
                                        constructor.getParameterTypes()));
    }

    public static final class MetaDataInner {
```

```
        private String name;

        private Integer age;

        @AnnotationTest({"construct"})
        public MetaDataInner(@AnnotationTest({"param1"}) String name,
                             @AnnotationTest({"param2"}) Integer age) {
            this.name = name;
            this.age = age;
        }

        public String getName() {
            return name;
        }

        public void setName(String name) {
            this.name = name;
        }

        public Integer getAge() {
            return age;
        }

        public void setAge(Integer age) {
            this.age = age;
        }
    }
}
```

执行结果如下：

```
[class java.lang.String, class java.lang.Integer]
```

6. getParameterCount()

返回此构造函数的入参数量。

7. getParameterAnnotations()

返回此构造函数的入参注解信息，代码如下：

```
/第 6 章/two/ConstructorTest.java
public class ConstructorTest {
    public static void main(String[] args) throws Exception {
        Constructor<?>[] constructors =
                MetaDataInner.class.getDeclaredConstructors();
        Constructor<?> constructor = constructors[0];
        Annotation[][] annotations = constructor.getParameterAnnotations();
        for (Annotation[] annotation : annotations) {
            System.out.println(Arrays.toString(annotation));
        }
```

```
    }

    public static final class MetaDataInner {
        private String name;

        private Integer age;

        @AnnotationTest({"construct"})
        public MetaDataInner(@AnnotationTest({"param1"}) String name,
                             @AnnotationTest({"param2"}) Integer age) {
            this.name = name;
            this.age = age;
        }

        public String getName() {
            return name;
        }

        public void setName(String name) {
            this.name = name;
        }

        public Integer getAge() {
            return age;
        }

        public void setAge(Integer age) {
            this.age = age;
        }
    }
}
```

执行结果如下：

```
[@cn.kungreat.niobook.six.one.AnnotationTest({"param1"})]
[@cn.kungreat.niobook.six.one.AnnotationTest({"param2"})]
```

8. isVarArgs()

检查此构造函数入参是否是可变参数，返回 boolean 值。

9. setAccessible(boolean flag)

设置此构造函数的可访问标志。接收 boolean 入参，作为可访问标志。

10. newInstance(Object…initargs)

使用此构造函数来创建新的实例对象。接收 Object…入参，作为构造函数入参，代码如下：

```
/第 6 章/two/ConstructorTest.java
public class ConstructorTest {
```

```
    public static void main(String[ ] args) throws Exception {
        Constructor<?>[ ] constructors =
                MetaDataInner.class.getDeclaredConstructors();
        Constructor<?> constructor = constructors[0];
        Object object = constructor.newInstance("kungreat", 18);
        System.out.println(object);
    }

    public static final class MetaDataInner {
        private String name;

        private Integer age;

        @AnnotationTest({"construct"})
        public MetaDataInner(@AnnotationTest({"param1"}) String name,
                             @AnnotationTest({"param2"}) Integer age) {
            this.name = name;
            this.age = age;
        }

        public String getName() {
            return name;
        }

        public void setName(String name) {
            this.name = name;
        }

        public Integer getAge() {
            return age;
        }

        public void setAge(Integer age) {
            this.age = age;
        }
            @Override
        public String toString() {
            return "MetaDataInner{" +
                    "name='" + name + '\'' +
                    ", age=" + age +
                    '}';
        }
    }
}
```

执行结果如下：

```
MetaDataInner{name='kungreat', age=18}
```

14min

6.3 Field 类

Field 类提供了类单个字段的相关信息和访问权。本节基于官方文档介绍常用方法，核心方法配合代码示例以方便读者理解。

1. get(Object obj)

返回指定对象上此字段表示的值，如果此字段是静态字段，则可以为空。接收 Object 入参，作为指定对象。

2. getAnnotations()

返回此字段上的注解信息。

3. getType()

返回一个类对象，用于标识此字段的类型。

4. getDeclaringClass()

返回一个类对象，用于标识此字段所属类的类型。

5. getBoolean(Object obj)

返回指定对象上此字段表示的布尔值，如果此字段是静态字段，则可以为空。接收 Object 入参，作为指定对象。

6. getByte(Object obj)

返回指定对象上此字段表示的字节值，如果此字段是静态字段，则可以为空。接收 Object 入参，作为指定对象。

7. getChar(Object obj)

返回指定对象上此字段表示的字符值，如果此字段是静态字段，则可以为空。接收 Object 入参，作为指定对象。

8. getDouble(Object obj)

返回指定对象上此字段表示的双精度浮点值，如果此字段是静态字段，则可以为空。接收 Object 入参，作为指定对象。

9. getFloat(Object obj)

返回指定对象上此字段表示的单精度浮点值，如果此字段是静态字段，则可以为空。接收 Object 入参，作为指定对象。

10. getInt(Object obj)

返回指定对象上此字段表示的整数值，如果此字段是静态字段，则可以为空。接收 Object 入参，作为指定对象。

11. getLong(Object obj)

返回指定对象上此字段表示的长整数值，如果此字段是静态字段，则可以为空。接收Object入参，作为指定对象。

12. getModifiers()

返回此字段表示的语言修饰符。

13. getShort(Object obj)

返回指定对象上此字段表示的短整数值，如果此字段是静态字段，则可以为空。接收Object入参，作为指定对象。

14. isEnumConstant()

检查此字段是否是枚举类型，返回boolean值。

15. set(Object obj, Object value)

设置指定对象上此字段表示的值，如果此字段是静态字段，则可以为空。接收Object入参，作为指定对象，接收Object入参，作为字段表示的值。

16. setAccessible(boolean flag)

设置此字段的可访问标志。接收boolean入参，作为可访问标志。

17. setBoolean(Object obj, boolean z)

设置指定对象上此字段表示的值，如果此字段是静态字段，则可以为空。接收Object入参，作为指定对象，接收boolean入参，作为字段表示的值。

18. setByte(Object obj, byte b)

设置指定对象上此字段表示的值，如果此字段是静态字段，则可以为空。接收Object入参，作为指定对象，接收byte入参，作为字段表示的值。

6.4 Method类

Method类提供了类单个方法的相关信息和访问权。本节基于官方文档介绍常用方法，核心方法配合代码示例以方便读者理解。

1. getDeclaringClass()

返回此方法类的描述对象。

2. getReturnType()

返回此方法返回类型的描述对象。

3. getExceptionTypes()

返回此方法所抛出的异常类型。

4．getParameterCount()

返回此方法的入参数量。

5．getParameterTypes()

返回此方法的入参信息。

6．getParameterAnnotations()

返回此方法的入参注解信息。

7．getDeclaredAnnotations()

返回此方法的注解信息。

8．isVarArgs()

检查此构造函数入参是否是可变参数，返回 boolean 值。

9．setAccessible(boolean flag)

设置此构造函数的可访问标志。接收 boolean 入参，作为可访问标志。

10．getModifiers()

返回此字段表示的语言修饰符。

11．invoke(Object obj, Object…args)

使用指定的入参在指定的对象上调用此方法，并返回方法的执行结果。接收 Object 入参，作为指定的对象，接收 Object… 入参，作为方法的入参，代码如下：

```
//第 6 章/four/MethodTest.java
//import 第 6 章/one/MetaDataTest.MetaDataInner.java
public class MethodTest {

    public static void main(String[] args) throws Exception {
        MetaDataTest.MetaDataInner meta = new
                        MetaDataTest.MetaDataInner("kungreat",18);
        Method method = MetaDataTest.MetaDataInner.class.
                                getDeclaredMethod("getName");
        Method method1 = MetaDataTest.MetaDataInner.class.
                        getDeclaredMethod("setName", String.class);

        System.out.println(method);
        System.out.println(Arrays.toString(
                        method.getDeclaredAnnotations()));
        System.out.println(Arrays.deepToString
                        (method.getParameterAnnotations()));
        System.out.println(method.getParameterCount());
        System.out.println(Arrays.toString(method.getParameterTypes()));
        System.out.println(method.isVarArgs());
        System.out.println(method.getDeclaringClass());

        int modifiers = method.getModifiers();
```

```
        System.out.println(Modifier.isStatic(modifiers));
        System.out.println(Modifier.isPublic(modifiers));
        System.out.println(method.getReturnType());
        System.out.println(method1.getReturnType());

        System.out.println(method.invoke(meta));
        System.out.println(method1.invoke(meta, "new name"));
        System.out.println(method.invoke(meta));
    }
}
```

执行结果如下：

```
public java.lang.String cn.kungreat.niobook.six.one.MetaDataTest$MetaDataInner.getName()
[]
[]
0
[]
false
class cn.kungreat.niobook.six.one.MetaDataTest$MetaDataInner
false
true
class java.lang.String
void
kungreat
null
new name
```

小结

封装了底层类的描述信息，通过反射机制可以解决很多硬编码解决不了的问题，反射机制被称为框架基石。

习题

1. 判断题

(1) 通过反射机制可以创建私有构造器的对象。(　　)

(2) 通过反射机制可以获取语言权限修饰符。(　　)

(3) 反射机制并不破坏面向对象的编程概念。(　　)

2. 选择题

(1) 通过反射机制可以获取的类型对象有(　　)。(多选)

A. Constructor 类　　　　B. Field 类

C. Method 类 D. Annotation 类

(2) Constructor 类构造对象实例的方法是(　　)。(单选)

A. newInstance(Object…initargs) B. isSynthetic()

C. getModifiers() D. getExceptionTypes()

(3) Method 类通过反射调用方法的是(　　)。(单选)

A. getReturnType() B. invoke(Object obj, Object…args)

C. isVarArgs() D. getDefaultValue()

(4) Field 类获取对象数据的方法有(　　)。(多选)

A. get(Object obj) B. getChar(Object obj)

C. getLong(Object obj) D. getName()

3. 填空题

查看执行结果并补充代码，代码如下：

```
//第 6 章/answer/OneAnswer.java
public class OneAnswer {

    public static void main(String[] args) throws Exception {
        Constructor<?> constructor = OneData.class.getConstructor
                                    (String.class, Integer.class);
        Object oneData = constructor.newInstance(________,________);
        Field[] fields = OneData.class.getDeclaredFields();
        //输出默认值
        for (Field field : fields) {
            System.out.println(field.get(________));
        }
        //通过方法设置新的值
        Method[] methods = OneData.class.getDeclaredMethods();
        for (Method method : methods) {
            int count = method.getParameterCount();
            if(count > 0){
                Class<?>[] parameterTypes = method.getParameterTypes();
                for (Class<?> type : parameterTypes) {
                    if(type == String.class){
                        method.invoke(oneData,________);
                    } else if (type == Integer.class) {
                        method.invoke(oneData,________);
                    }
                }
            }
        }
        //输出新的值
        for (Field field : fields) {
            System.out.println(field.get(oneData));
        }
```

```
    }

    private static final class OneData{
        private String name;
        private Integer age;

        public OneData(String name, Integer age) {
            this.name = name;
            this.age = age;
        }

        public String getName() {
            return name;
        }

        public void setName(String name) {
            this.name = name;
        }

        public Integer getAge() {
            return age;
        }

        public void setAge(Integer age) {
            this.age = age;
        }
    }
}
```

执行结果如下：

```
kungreat
188
new name
666
```

第 7 章

CHAPTER 7

ClassLoader 类加载器

抽象类 ClassLoader 定义了类的加载方式。

40min

7.1 ClassLoader 抽象类

类加载器负责把数据加载进 JVM 内存中，类加载器应尝试查找类定义的数据。典型的策略是将名称转换为文件名，然后从文件系统中读取该名称的数据。

7.1.1 基本介绍

JDK 默认提供了以下 3 个内置的类加载器：

(1) Bootstrap 是虚拟机的内置类加载器，通常表示为空并且没有上级类加载器。

(2) PlatformClassLoader 是平台类加载器，负责装入平台类。它的上级是虚拟机的内置类加载器。

(3) AppClassLoader 是系统类加载器，称为应用程序类加载器。它的上级是平台类加载器。

示例代码如下：

```
//第 7 章/one/ClassLoadTest.java
public class ClassLoadTest {

    public static void main(String[] args) throws Exception {
        System.out.println(String.class.getClassLoader());
        ClassLoader loader = ClassLoader.getSystemClassLoader();
        while (loader != null) {
            System.out.println(loader);
            loader = loader.getParent();
        }
    }
}
```

执行结果如下：

```
null
jdk.internal.loader.ClassLoaders$AppClassLoader@63947c6b
jdk.internal.loader.ClassLoaders$PlatformClassLoader@776ec8df
```

7.1.2 自定义加载器

简单地自定义一个本地文件资源类加载器，代码如下：

```java
//第7章/one/ClassLoadTest.java
public class ClassLoadTest {
public static void main(String[] args) throws Exception {
    //将类加载器注册为并行模式
        System.out.println(FileClassLoad.registerAsParallelCapable());
        FileClassLoad fileClassLoad = new FileClassLoad(new File(
                "C:\\Users\\mydre\\Desktop\\another\\out\\production
                        \\another\\com\\kungreat\\PackAgeTest.class"));
        Class<?> aClass =
    fileClassLoad.loadClass("com.kungreat.PackAgeTest");
        Method main = aClass.getDeclaredMethod("main", String[].class);
        main.invoke(null, (Object) new String[]{"FileClassLoad",
                                                    " - loadClass"});
        System.out.println(aClass.getClassLoader());
    }

    private static final class FileClassLoad extends ClassLoader {

        private final File file;

        private FileClassLoad(File file) {
            this.file = file;
        }
    //当上级类加载器没有找到资源时会回调此方法
        public Class<?> findClass(String name) {
            byte[] b = loadClassData();
            return defineClass(name, b, 0, b.length);
        }

        private byte[] loadClassData() {
            try (FileInputStream stream = new FileInputStream(file)) {
                return stream.readAllBytes();
            } catch (Exception e) {
                throw new RuntimeException(e.getMessage());
            }
        }

        public static boolean registerAsParallelCapable() {
            return ClassLoader.registerAsParallelCapable();
        }
    }
}
```

执行结果如下：

```
true
[FileClassLoad, - loadClass]
cn.kungreat.niobook.seven.two.ClassLoadTest$FileClassLoad@4eec7777
```

PackAgeTest 类的源代码，如图 7-1 所示。

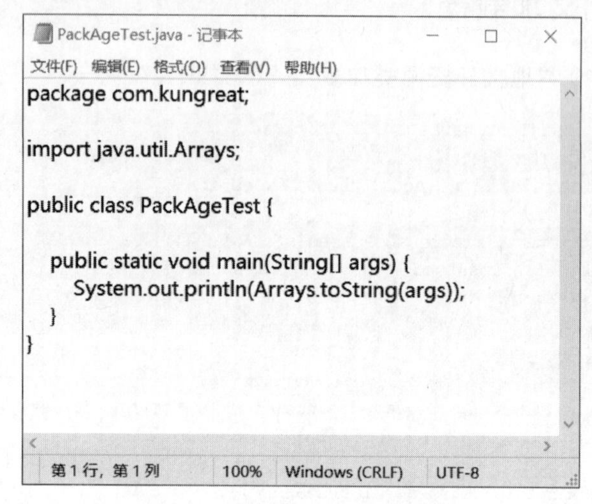

图 7-1 PackAgeTest 类的源代码

类加载器的流程采用的是由上至下的委托模式，如图 7-2 所示。

7.1.3 URLClassLoader 类

此类加载器用于从 JAR 文件和目录中搜索并加载类资源，代码如下：

```
//第 7 章/one/ClassLoadTest.java
public class ClassLoadTest {
    public static void main(String[] args) throws Exception {
        URLClassLoader urlClassLoader = new URLClassLoader(new URL[]{
                new URL("file:C:\\Users\\mydre\\Desktop
                            \\another\\out\\production\\another\\")});
        Class<?> aClass =
    urlClassLoader.loadClass("com.kungreat.PackAgeTest");
        Method main = aClass.getDeclaredMethod("main", String[].class);
        main.invoke(null, (Object) new String[]
                                                {"FileClassLoad"," - loadClass"});
        System.out.println(aClass.getClassLoader());
    }
}
```

执行结果如下：

```
true
[FileClassLoad, - loadClass]
java.net.urlClassLoader@3b07d329
```

```
protected Class<?> loadClass(String name, boolean resolve)
    throws ClassNotFoundException {
    synchronized (getClassLoadingLock(name)) {
        //首先检查类是否已经加载
        Class<?> c = findLoadedClass(name);
        if (c == null) {
            long t0 = System.nanoTime();
            try {
                if (parent != null) {
                    //委托上级类加载器
                    c = parent.loadClass(name, false);
                } else {
                    //当上级类加载器不存在时
                    c = findBootstrapClassOrNull(name);
                }
            } catch (ClassNotFoundException e) {
                //当没有找到类时
            }

            if (c == null) {
                long t1 = System.nanoTime();
                //当没有找到类时回调子类方法
                c = findClass(name);

                //记录统计数据
                PerfCounter.getParentDelegationTime().addTime(t1 - t0);
                PerfCounter.getFindClassTime().addElapsedTimeFrom(t1);
                PerfCounter.getFindClasses().increment();
            }
        }
        if (resolve) {
            resolveClass(c);
        }
        return c;
    }
}
```

图 7-2 类加载器的流程

注意：如果一定要使用自定义类加载器，则推荐使用 URLClassLoader 类加载器。

7.2 Annotation 接口

所有的注解类都自动实现了 Annotation 接口。

7.2.1 注解的限制

(1) 一个注解不能继承另一个注解。

(2) 注解声明的属性都不能带入参。

(3) 注解的属性类型只能是基本类型、String 类、Class 类、枚举类型、其他注解类型、合法类型的数组。

7.2.2 内置注解

本节介绍官方提供的常用内置注解。

1. @Target

用于指定注解可以出现的位置，此注解只能用在其他注解类型的声明上。入参是ElementType枚举类型的常量，见表7-1。

表7-1 ElementType枚举类型的常量

枚举常量	描述信息
ANNOTATION_TYPE	注解类型
CONSTRUCTOR	构造器
FIELD	字段
LOCAL_VARIABLE	局部变量
METHOD	方法
MODULE	模块
PACKAGE	包
PARAMETER	入参
RECORD_COMPONENT	记录类
TYPE	类、接口、枚举
TYPE_PARAMETER	类型入参
TYPE_USE	类型使用

2. @Retention

用于指定注解的生命周期，此注解只能用在其他注解类型的声明上。入参是RetentionPolicy枚举类型的常量。

3. @Inherited

用于使子类也能直接发现超类上的注解，此注解只能用在其他注解类型的声明上。

4. @Override

用于强制检查方法的重写，此注解只能用在方法声明上。

5. @Deprecated

用于指出当前的声明是过时的，此注解可以出现在多个位置。

6. @SuppressWarnings

用于抑制编译器可能产生的警告信息，入参是字符串数组。

7.2.3 自定义注解

使用自定义注解，代码如下：

```java
//第 7 章/two/MyAnnotation.java
@Target({TYPE, FIELD, METHOD, PARAMETER, CONSTRUCTOR, LOCAL_VARIABLE, MODULE})
@Retention(RetentionPolicy.RUNTIME)
@Inherited
@Repeatable(MyAnnotationRepeat.class)
public @interface MyAnnotation {
    String value();
    int index();
    String name() default "kungreat";
}

//第 7 章/two/MyAnnotationRepeat.java
@Target({TYPE, FIELD, METHOD, PARAMETER, CONSTRUCTOR, LOCAL_VARIABLE, MODULE})
@Retention(RetentionPolicy.RUNTIME)
@Inherited
public @interface MyAnnotationRepeat {
    MyAnnotation[] value();
}
```

测试自定义注解，代码如下：

```java
//第 7 章/two/AnnotationParent.java
@MyAnnotation(value = "one",index = 18)
@MyAnnotation(value = "two",index = 28)
public class AnnotationParent {

}

//第 7 章/two/AnnotationTest.java
public class AnnotationTest extends AnnotationParent{

    public static void main(String[] args) {
        MyAnnotation annotation = AnnotationTest.class.
                                getAnnotation(MyAnnotation.class);
        System.out.println(annotation);

        MyAnnotationRepeat annotationRepeat = AnnotationTest.class.
                            getAnnotation(MyAnnotationRepeat.class);
        System.out.println(annotationRepeat);
    }

}
```

执行结果如下：

```
null
@cn.kungreat.niobook.seven.two.MyAnnotationRepeat({@cn.kungreat.niobook.seven.two.
MyAnnotation(name = "kungreat", value = "one", index = 18),
@cn.kungreat.niobook.seven.two.MyAnnotation(name = "kungreat", value = "two", index = 28)})
```

注意：注解本身只是一种标签或标记，不具备任何功能实现。

小结

通过本章的学习需要理解类加载器的概念，并可以实现自定义的类加载器。也需要理解注解的概念，后续将在自研框架中使用注解并配合反射机制实现方法的动态调用。

习题

1. 判断题

(1) 每个 Class 类对象都包含其类加载器的引用。(　　)

(2) ClassLoader 类加载器使用委托机制由上至下加载资源。(　　)

(3) 注解本身只是一种标签或标记，不具备任何功能实现。(　　)

2. 选择题

(1) 平台内置注解类有(　　)。(多选)

A. @Inherited 类　　B. @Override 类

C. @Target 类　　D. @Deprecated 类

(2) 平台内置注解有强制方法重写检查的类是(　　)。(单选)

A. @Inherited 类　　B. @Override 类

C. @Target 类　　D. @Deprecated 类

(3) ClassLoader 类加载器平台默认提供的有(　　)。(多选)

A. AppClassLoader 类　　B. PlatformClassLoader 类

C. URLClassLoader 类　　D. Object 类

(4) 用于指定注解生命周期的类是(　　)。(单选)

A. @Retention 类　　B. @Override 类

C. @Target 类　　D. @Deprecated 类

3. 填空题

PackAgeTest 类的源代码如下：

```
public class PackAgeTest {

    public static void main(String[ ] args) {
        System.out.println(Arrays.toString(args));
    }
}
```

(1) 查看执行结果并补充代码,代码如下:

```
//第 7 章/answer/ClassLoadAnswer.java
public class ClassLoadAnswer {
    public static void main(String[] args) throws Exception {
        System.out.println(FileClassLoad.registerAsParallelCapable());
        FileClassLoad fileClassLoad = new FileClassLoad(
                    new File("C:\\Users\\mydre\\Desktop\\another\\out\\
    production\\another\\com\\kungreat\\PackAgeTest.class"));
        Class<?> aClass = fileClassLoad.loadClass
                                    ("________.________.________");
        Method main = aClass.getDeclaredMethod("main", String[].class);
        main.invoke(null, (Object) new String[]
                                {"________", "________"});
        System.out.println(aClass.getClassLoader());
    }

    private static final class FileClassLoad extends ClassLoader {

        private final File file;

        private FileClassLoad(File file) {
            this.file = file;
        }

        public Class<?> ________(String name) {
            byte[] b = loadClassData();
            return defineClass(name, b, 0, b.length);
        }

        private byte[] loadClassData() {
            try (FileInputStream stream = new FileInputStream(file)) {
                return stream.readAllBytes();
            } catch (Exception e) {
                throw new RuntimeException(e.getMessage());
            }
        }

        public static boolean registerAsParallelCapable() {
            return ClassLoader.registerAsParallelCapable();
        }
    }
}
```

执行结果如下:

```
true
[FileClassLoad, - loadClass]
cn.kungreat.niobook.seven.answer.ClassLoadAnswer$FileClassLoad@4eec7777
```

（2）查看执行结果并补充代码，代码如下：

```
//第 7 章/answer/TwoAnswer.java
public class TwoAnswer {
    public static void main(String[] args) throws Exception {
        URLClassLoader urlClassLoader = new URLClassLoader(
                new URL[]{ new URL("file:C:\\Users\\mydre\\Desktop\\another
                                \\out\\production\\another\\")});
        Class<?> aClass = urlClassLoader.loadClass
                                ("________.________.________");
        Method main = aClass.getDeclaredMethod("main", String[].class);
        main.invoke(null, (Object) new String[]
                                {"________","________________"});
        System.out.println(aClass.getClassLoader());
    }

}
```

执行结果如下：

```
[FileClassLoad, - loadClass]
java.net.urlClassLoader@4eec7777
```

第8章 网络基础

CHAPTER 8

本章介绍网络基础知识，如 IP 地址、网络接口、用户数据报协议（User Datagram Protocol，UDP）。

8.1 InetAddress 类

此类表示因特网协议中的 IP 地址，其中 IPv4 占用 4 字节，即 32 位，IPv6 占用 16 字节，即 128 位。这是一个较低级别的协议，在此基础上构建了 UDP 和传输控制协议（Transmission Control Protocol，TCP）。

本节基于官方文档介绍常用方法，核心方法配合代码示例以方便读者理解。

8.1.1 核心方法

1. getAllByName(String host)

静态方法，根据指定的域名查询并返回其所属 IP 地址的数组。接收 String 入参，作为指定的域名，代码如下：

```
//第 8 章/one/InetAddressTest.java
public class InetAddressTest {

    public static void main(String[] args) throws Exception {
        InetAddress[] allByName = InetAddress.getAllByName
                                    ("www.kungreat.cn");
        System.out.println(Arrays.toString(allByName));
    }
}
```

执行结果如下：

```
[www.kungreat.cn/39.100.107.108]
```

2. getByAddress(byte[] addr)

静态方法，返回给定原始 IP 地址的 InetAddress 对象。接收 byte[]入参，作为原始 IP

地址，代码如下：

```
//第 8 章/one/InetAddressTest.java
public class InetAddressTest {

    public static void main(String[] args) throws Exception {
        InetAddress byIp = InetAddress.getByAddress(new
                                              byte[]{39,100,107,108});
        System.out.println(byIp);
        System.out.println(byIp.getCanonicalHostName());
        System.out.println(byIp.getHostAddress());
        System.out.println(byIp.getHostName());
    }
}
```

执行结果如下：

```
/39.100.107.108
39.100.107.108
39.100.107.108
39.100.107.108
```

3. getByName(String host)

静态方法，根据指定的域名查询并返回其中的一个 IP 地址。接收 String 入参，作为指定的域名，代码如下：

```
public class InetAddressTest {

    public static void main(String[] args) throws Exception {
        InetAddress byIp = InetAddress.getByName("www.kungreat.cn");
        System.out.println(byIp);
        System.out.println(byIp.getCanonicalHostName());
        System.out.println(byIp.getHostAddress());
        System.out.println(byIp.getHostName());
    }
}
```

执行结果如下：

```
www.kungreat.cn/39.100.107.108
39.100.107.108
39.100.107.108
www.kungreat.cn
```

4. getAddress()

返回当前对象的原始 IP 地址。

5. getCanonicalHostName()

返回当前对象 IP 地址的标准域名。

6. getHostAddress()

返回当前对象 IP 地址的字符串。

7. getHostName()

返回当前对象 IP 地址的主机名。

8. getLocalHost()

返回本地主机其中的一个 IP 地址所表示的 InetAddress 对象。

9. getLoopbackAddress()

返回本地主机回环 IP 地址所表示的 InetAddress 对象。

10. isAnyLocalAddress()

检查当前对象 IP 地址是否是通配符类型,返回 boolean 值,代码如下:

```java
//第 8 章/one/InetAddressTest.java
public class InetAddressTest {

    public static void main(String[] args) throws Exception {
        InetAddress byIp = InetAddress.getByName("0.0.0.0");
        System.out.println(byIp);
        System.out.println(byIp.isAnyLocalAddress());
        System.out.println(byIp.isLinkLocalAddress());
        System.out.println(byIp.isLoopbackAddress());
    }
}
```

执行结果如下:

```
/0.0.0.0
true
false
false
```

11. isLinkLocalAddress()

检查当前对象 IP 地址是否是链接本地地址类型,返回 boolean 值。

12. isLoopbackAddress()

检查当前对象 IP 地址是否是本地回环类型,返回 boolean 值。

13. isMCGlobal()

检查当前对象 IP 地址是否是多点广播地址类型,并且具有全局作用域,返回 boolean 值。

14. isMCLinkLocal()

检查当前对象 IP 地址是否是多点广播地址类型,并且具有链接作用域,返回 boolean 值。

15. isMCNodeLocal()

检查当前对象 IP 地址是否是多点广播地址类型,并且具有节点作用域,返回

boolean 值。

16. isMCOrgLocal()

检查当前对象 IP 地址是否是多点广播地址类型，并且具有组织范围，返回 boolean 值。

17. isMCSiteLocal()

检查当前对象 IP 地址是否是多点广播地址类型，并且具有站点作用域，返回 boolean 值。

18. isMulticastAddress()

检查当前对象 IP 地址是否是多点广播地址类型，返回 boolean 值。

19. isReachable(int timeout)

检查当前对象 IP 地址是否可以正常连接，返回 boolean 值。接收 int 入参，作为最长超时时间毫秒数。

20. isSiteLocalAddress()

检查当前对象 IP 地址是否是本地站点地址类型，返回 boolean 值。

8.1.2 DNS 服务

使用域名服务时会调用本地 DNS 服务来查询 IP 地址，代码如下：

```
//第 8 章/one/InetAddressTest.java
public class InetAddressTest {

    public static void main(String[] args) throws Exception {
        InetAddress[] allByName = InetAddress.getAllByName
                                            ("www.kungreat.cn");
        System.out.println(Arrays.toString(allByName));
    }
}
```

执行结果如下：

```
[www.kungreat.cn/39.100.107.108]
```

使用 Wireshark 软件监控网络流量，如图 8-1 所示。

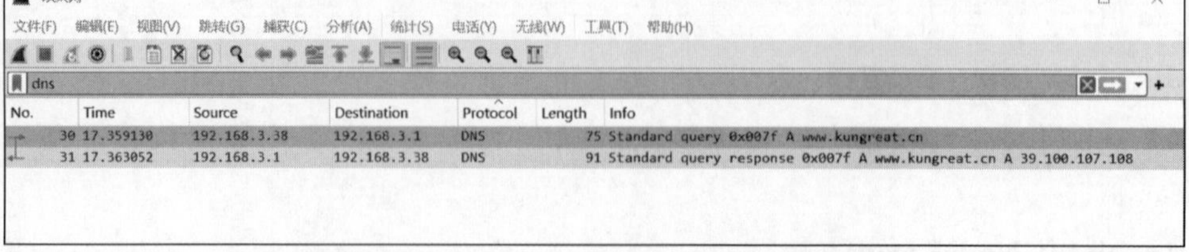

图 8-1 Wireshark 软件监控网络流量

8.2 NetworkInterface 类

13min

表示由名称组成的网络接口,以及分配给此接口的 IP 地址列表。本节基于官方文档介绍常用方法,核心方法配合代码示例以方便读者理解。

1. getByIndex(int index)

静态方法,返回指定索引的网络接口。接收 int 入参,作为指定索引。

2. getByInetAddress(InetAddress addr)

静态方法,返回指定 IP 地址的网络接口。接收 InetAddress 入参,作为指定 IP 地址。

3. getByName(String name)

静态方法,返回指定名称的网络接口。接收 String 入参,作为指定名称。

4. getNetworkInterfaces()

静态方法,返回本地主机上的所有网络接口,代码如下:

```
//第 8 章/two/NetworkInterfaceTest.java
public class NetworkInterfaceTest {

   public static void main(String[] args) throws Exception {
      Enumeration<NetworkInterface> interfaces = NetworkInterface.
                                                 getNetworkInterfaces();
      while (interfaces.hasMoreElements()) {
         NetworkInterface nexted = interfaces.nextElement();
      //网络接口是否已启动并正在运行
         if(nexted.isUp()){
            System.out.println(nexted);
         }
      }
   }
}
```

执行结果如下:

```
name:lo (Software Loopback Interface 1)
name:eth2 (VMware Virtual Ethernet Adapter for VMnet8)
name:eth4 (Realtek Gaming 2.5GbE Family Controller)
name:eth5 (VMware Virtual Ethernet Adapter for VMnet1)
```

本地的网络设备如图 8-2 所示。

5. getDisplayName()

返回当前网络接口的显示名称。

6. getHardwareAddress()

返回当前网络接口的 MAC 地址。

图 8-2　本地的网络设备

7．getIndex()

返回当前网络接口的索引。

8．getInetAddresses()

返回当前网络接口的 IP 地址列表，代码如下：

```
//第 8 章/two/NetworkInterfaceTest.java
public class NetworkInterfaceTest {

    public static void main(String[] args) throws Exception {
        Enumeration<NetworkInterface> interfaces = NetworkInterface.
                                                    getNetworkInterfaces();
        while (interfaces.hasMoreElements()) {
            NetworkInterface nexted = interfaces.nextElement();
            if(nexted.isUp()){
                Enumeration<InetAddress> inetAddresses = nexted.
                                                    getInetAddresses();
                while (inetAddresses.hasMoreElements()){
                    System.out.println(inetAddresses.nextElement());
                }
            }
        }
    }
}
```

执行结果如下：

```
/127.0.0.1
/0:0:0:0:0:0:0:1
/192.168.80.1
/fe80:0:0:0:376f:c66b:81a8:f066%eth2
/192.168.3.38
/fe80:0:0:0:f0be:c665:e755:f71d%eth4
/192.168.44.1
/fe80:0:0:0:c714:92ef:5ac6:6388%eth5
```

9. getInterfaceAddresses()

返回当前网络接口的所有 InterfaceAddresses 信息列表。

10. getMTU()

返回当前网络接口的最大传输单元。

11. getName()

返回当前网络接口的名称。

12. getParent()

如果当前网络接口是子接口,则返回它的上级网络接口,否则返回空。

13. getSubInterfaces()

返回当前网络接口的所有子接口列表。

14. isLoopback()

检查当前网络接口是否是本地回环接口,返回 boolean 值。

15. isPointToPoint()

检查当前网络接口是否是点到点接口,返回 boolean 值。

16. isUp()

检查当前网络接口是否已启动并正在运行,返回 boolean 值。

17. isVirtual()

检查当前网络接口是否是虚拟网络接口,返回 boolean 值。

18. supportsMulticast()

检查当前网络接口是否支持多点广播,返回 boolean 值。

8.3 URI 类

URI 类表示统一资源标识符。提供用于从组件创建 URI 实例的构造器、访问 URI 实例各种组件的方法及规范化解析和传递 URI 实例的方法。

8.3.1 构造器

URI 构造器见表 8-1。

表 8-1 URI 构造器

构 造 器	描 述
URI(String str)	构造新的对象,指定解析的字符串
URI (String scheme, String ssp, String fragment)	构造新的对象,指定协议,指定协议特定部分,指定片段

续表

构　造　器	描　　述
URI(String scheme, String userInfo, String host, int port, String path, String query, String fragment)	构造新的对象，指定协议，指定用户名和授权，指定主机名称，指定绑定的端口，指定路径，指定查询条件，指定片段
URI(String scheme, String host, String path, String fragment)	构造新的对象，指定协议，指定主机名称，指定路径，指定片段
URI(String scheme, String authority, String path, String query, String fragment)	构造新的对象，指定协议，指定授权，指定路径，指定查询条件，指定片段

8.3.2 常用方法

本节基于官方文档介绍常用方法，核心方法配合代码示例以方便读者理解。

1. getAuthority()

返回当前对象已解码的授权信息。

2. getFragment()

返回当前对象已解码的片段信息。

3. getHost()

返回当前对象已解码的主机名称信息。

4. getPath()

返回当前对象已解码的路径信息。

5. getPort()

返回当前对象的端口号。

6. getQuery()

返回当前对象已解码的查询条件信息。

7. getRawAuthority()

返回当前对象原始的授权信息。

8. getRawFragment()

返回当前对象原始的片段信息。

9. getRawPath()

返回当前对象原始的路径信息。

10. getRawQuery()

返回当前对象原始的查询条件信息。

11. getRawSchemeSpecificPart()

返回当前对象原始的协议特定部分信息。

12. getRawUserInfo()

返回当前对象原始的用户名和授权信息。

13. getScheme()

返回当前对象的协议信息。

14. getSchemeSpecificPart()

返回当前对象已解码的协议特定部分信息。

15. getUserInfo()

返回当前对象已解码的用户名和授权信息。

16. isAbsolute()

检查当前对象是否是绝对的数据类型，返回 boolean 值，代码如下：

```
//第 8 章/three/UriTest.java
public class UriTest {

    public static void main(String[] args) throws Throwable {
        URI uri = new URI("https://www.example.com:443/languages
    /java/sample/a/index.html?name=kungreat&age=18#28");
        System.out.println(uri.getScheme());
        System.out.println(uri.getSchemeSpecificPart());
        System.out.println(uri.getAuthority());
        System.out.println(uri.getUserInfo());
        System.out.println(uri.getHost());
        System.out.println(uri.getPort());
        System.out.println(uri.getPath());
        System.out.println(uri.getQuery());
        System.out.println(uri.getFragment());

        URI uri1 = new URI("https://www.kungreat.cn/?
                    name=%E5%88%98%E5%A4%A7%E8%83%96");
        System.out.println(uri1.getQuery());
        System.out.println(uri1.getRawQuery());
        System.out.println(uri1.isAbsolute());
        System.out.println(uri1.isOpaque());
    }
}
```

执行结果如下：

```
https
//www.example.com:443/languages/java/sample/a/index.html?name=kungreat&age=18
www.example.com:443
```

```
null
www.example.com
443
/languages/java/sample/a/index.html
name = kungreat&age = 18
28
name = 刘大胖
name = %E5%88%98%E5%A4%A7%E8%83%96
true
false
```

17. isOpaque()

检查当前对象是否是透明的数据类型，返回 boolean 值，代码如下：

```
//第 8 章/three/UriTest.java
public class UriTest {

    public static void main(String[] args) throws Throwable {
        URI uri1 = new URI("mailto://java-net@www.example.com");
        System.out.println(uri1.getUserInfo());
        System.out.println(uri1.getHost());
        System.out.println(uri1.getAuthority());
        System.out.println(uri1.isAbsolute());
        System.out.println(uri1.isOpaque());
    }
}
```

执行结果如下：

```
java-net
www.example.com
java-net@www.example.com
true
false
```

16min

8.4 URL 类

URL 类表示统一资源定位符，在 URI 类的基础上额外提供了获取资源数据流的方法。

8.4.1 构造器

URL 构造器见表 8-2。

表 8-2 URL 构造器

构 造 器	描 述
URL(String spec)	构造新的对象,指定解析的字符串
URL(String protocol, String host, int port, String file)	构造新的对象,指定协议,指定主机名称,指定绑定的端口,指定文件
URL(String protocol, String host, int port, String file,URLStreamHandler handler)	构造新的对象,指定协议,指定主机名称,指定绑定的端口,指定文件,指定协议的流处理器
URL(String protocol, String host, String file)	构造新的对象,指定协议,指定主机名称,指定文件
URL(URL context, String spec)	构造新的对象,指定解析的上下文,指定解析的字符串
URL(URL context, String spec, URLStreamHandler handler)	构造新的对象,指定解析的上下文,指定解析的字符串,指定协议的流处理器

8.4.2 常用方法

本节基于官方文档介绍常用方法,核心方法配合代码示例以方便读者理解。

1. getAuthority()

返回当前对象已解码的授权信息。

2. getFile()

返回当前对象的文件信息。

3. getHost()

返回当前对象已解码的主机名称信息。

4. getPath()

返回当前对象已解码的路径信息。

5. getPort()

返回当前对象的端口号。

6. getProtocol()

返回当前对象的协议信息。

7. getQuery()

返回当前对象已解码的查询条件信息。

8. getRef()

返回当前对象已解码的锚点信息。

9. getUserInfo()

返回当前对象已解码的用户名和授权信息。

10. openConnection()

返回一个 URLConnection 实例，它表示与当前对象关联的连接信息。

11. toURI()

返回与当前对象等价的 URI 对象。

12. openStream()

返回一个 InputStream 实例，它表示与当前对象关联的连接信息内的字节流数据，代码如下：

```java
//第 8 章/four/URLTest.java
public class URLTest {
    public static void main(String[] args) throws Exception {
        URL url = new URL("https://www.kungreat.cn/java#88");
        System.out.println(url.getFile());
        System.out.println(url.getPath());
        System.out.println(url.getRef());
        try (InputStream inputStream = url.openStream()) {
            byte[] bytes = inputStream.readAllBytes();
            System.out.println(new String(bytes));
        } catch (Exception e) {
            e.printStackTrace();
        }

        URL url1 = new URL("file:D:\\temp\\src\\test.txt");
        System.out.println(url1.getFile());
        try (InputStream inputStream = url1.openStream()) {
            byte[] bytes = inputStream.readAllBytes();
            System.out.println(new String(bytes));
        } catch (Exception e) {
            e.printStackTrace();
        }

    }
}
```

执行结果如下：

```
/java
/java
88
<!doctype html><html lang="zh-CN"><head><meta charset="utf-8"/><meta http-equiv="X-UA-Compatible" content="IE=edge,chrome=1"/><meta name="keywords" content="kungreat,Java多线程并发体系实战，刘宁萌"/><link rel="icon" href="/favicon.ico"/><meta name="viewport" content="width=device-width,initial-scale=1,minimum-scale=1,maximum-scale=1,user-scalable=yes"/><meta name="theme-color" content="#000000"/><meta name="description" content="cpdogbbs"/><title>cpdogbbs</title><script defer="defer" src="/static/js/main.d1a53ee8.js"></script><link href="/static/css/main.899b0ab0.css" rel="stylesheet"></head><body><noscript>You need to enable JavaScript to
```

```
run this app.</noscript><div id="root"></div><audio src="/clear.mp3" id="clearMp3"
style="display:none"></audio></body></html>

D:/temp/src/test.txt
helloworld
endhelloworld
endhelloworld
end
```

文件内容如图 8-3 所示。

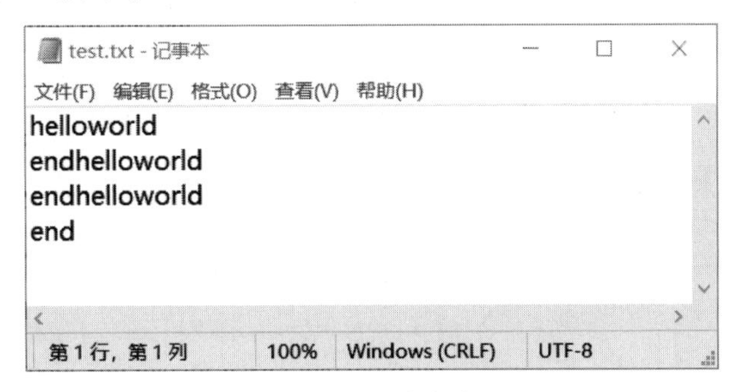

图 8-3 文件内容

8.5 JarURLConnection 抽象类

通过此类实例对象可以获取 JAR 包内的所有资源数据。

8.5.1 协议规则

常见协议规则如下。

(1) JAR 包：jar:file://www.home.com/duke/duke.jar!/。

(2) JAR 目录：jar:file://www.home.com/duke/duke.jar!/COM/foo/。

(3) JAR 文件：jar:file://www.home.com/duke/duke.jar!/COM/foo/Quux.class。

8.5.2 常用方法

本节基于官方文档介绍常用方法，核心方法配合代码示例以方便读者理解。

1. getAttributes()

如果当前对象指向的是 JAR 文件条目，则返回属性对象，否则返回空。

2. getMainAttributes()

返回当前对象 JAR 文件的主属性，代码如下：

```
//第 8 章/five/JarURLConnectionTest.java
public class JarURLConnectionTest {

    public static void main(String[] args) throws Exception {
        URL url = new URL("jar:file:
              C:/Users/mydre/IdeaProjects/niobook/temp/cpdog-1.0.jar!/");
        JarURLConnection jarConnection = (JarURLConnection)
                                          url.openConnection();
        System.out.println(jarConnection.getAttributes());
        System.out.println(jarConnection.getMainAttributes().entrySet());
    }
}
```

执行结果如下：

```
null
[Manifest-Version=1.0, Created-By=Maven Archiver 3.4.0, Build-Jdk-Spec=17, Class-Path=lib/HikariCP-5.0.1.jar lib/slf4j-api-2.0.0-alpha1.jar lib/logback-classic-1.3.0-alpha5.jar lib/logback-core-1.3.0-alpha5.jar lib/javax.mail-1.6.2.jar lib/activation-1.1.jar lib/checker-framework-1.7.0.jar lib/mysql-connector-java-8.0.22.jar lib/protobuf-java-3.11.4.jar lib/jackson-databind-2.12.4.jar lib/jackson-annotations-2.12.4.jar lib/jackson-core-2.12.4.jar, Main-Class=cn.kungreat.boot.CpdogMain]
```

3. getEntryName()

返回当前对象连接条目的名称。

4. getManifest()

返回当前对象的清单及其关联属性，如果没有，则为空，代码如下：

```
//第 8 章/five/JarURLConnectionTest.java
public class JarURLConnectionTest {

    public static void main(String[] args) throws Exception {
        URL url = new URL("jar:file:
              C:/Users/mydre/IdeaProjects/niobook/temp/cpdog-1.0.jar!/");
        JarURLConnection jarConnection = (JarURLConnection)
                                          url.openConnection();
        System.out.println(jarConnection.getManifest()
                                  .getMainAttributes().entrySet());
    }
}
```

执行结果如下：

```
[Manifest-Version=1.0, Created-By=Maven Archiver 3.4.0, Build-Jdk-Spec=17, Class-Path=lib/HikariCP-5.0.1.jar lib/slf4j-api-2.0.0-alpha1.jar lib/logback-classic-1.3.0-alpha5.jar lib/logback-core-1.3.0-alpha5.jar lib/javax.mail-1.6.2.jar lib/activation-1.1.jar lib/checker-framework-1.7.0.jar lib/mysql-connector-java-8.0.22.jar lib/protobuf-java-3.11.4.jar lib/jackson-databind-2.12.4.jar lib/jackson-annotations-2.12.4.jar lib/jackson-core-2.12.4.jar, Main-Class=cn.kungreat.boot.CpdogMain]
```

5. getJarFile()

返回当前对象的 JarFile 实例,代码如下:

```
//第 8 章/five/JarURLConnectionTest.java
public class JarURLConnectionTest {

    public static void main(String[] args) throws Exception {
        URL url = new URL("jar:file:
                C:/Users/mydre/IdeaProjects/niobook/temp/cpdog-1.0.jar!/");
        JarURLConnection jarConnection = (JarURLConnection)
                                                url.openConnection();
        JarFile jarFile = jarConnection.getJarFile();
        Enumeration<JarEntry> entries = jarFile.entries();
        while (entries.hasMoreElements()) {
            JarEntry jarEntry = entries.nextElement();
            System.out.println(jarEntry.getRealName());
        }
    }
}
```

执行结果如下:

```
META-INF/MANIFEST.MF
META-INF/
cn/
cn/kungreat/
cn/kungreat/boot/
cn/kungreat/boot/an/
cn/kungreat/boot/em/
cn/kungreat/boot/enents/
cn/kungreat/boot/exp/
cn/kungreat/boot/filter/
cn/kungreat/boot/handler/
cn/kungreat/boot/impl/
cn/kungreat/boot/jb/
cn/kungreat/boot/services/
cn/kungreat/boot/tls/
cn/kungreat/boot/utils/
META-INF/maven/
META-INF/maven/cn.kungreat/
META-INF/maven/cn.kungreat/cpdog/
7740639_www.kungreat.cn.jks
cn/kungreat/boot/handler/WebSocketConvertData$ReceiveObj.class
cn/kungreat/boot/NioBossServerSocket.class
logback.xml
temp.txt
base.txt
cn/kungreat/boot/handler/WebSocketConvertData$WebSocketData.class
mysql.txt
```

```
CleanupInstructions.txt
cn/kungreat/boot/handler/WebSocketConvertData.class
cn/kungreat/boot/services/FileService.class
cn/kungreat/boot/tls/CpDogSSLContext.class
cn/kungreat/boot/an/CpdogController.class
cn/kungreat/boot/services/ChartsService.class
cn/kungreat/boot/an/CpdogEvent.class
cn/kungreat/boot/services/LoginService.class
www.kungreat.cn.jks
cn/kungreat/boot/ChannelInHandler.class
cn/kungreat/boot/handler/WebSocketProtocolHandler.class
cn/kungreat/boot/ChannelOutHandler.class
cn/kungreat/boot/impl/BossThreadGroup.class
cn/kungreat/boot/tls/InitLinkedList.class
cn/kungreat/boot/ChannelProtocolHandler.class
cn/kungreat/boot/impl/ChooseWorkServerImpl.class
cn/kungreat/boot/tls/ShakeHands$CpdogThread.class
cn/kungreat/boot/ChooseWorkServer.class
cn/kungreat/boot/impl/NioBossServerSocketImpl$BossRunnable.class
cn/kungreat/boot/tls/ShakeHands$CpdogThreadFactory.class
cn/kungreat/boot/ConvertDataInHandler.class
cn/kungreat/boot/impl/NioBossServerSocketImpl$TLSRunnable.class
cn/kungreat/boot/tls/ShakeHands$DiscardPolicy.class
cn/kungreat/boot/tls/ShakeHands.class
cn/kungreat/boot/ConvertDataOutHandler.class
cn/kungreat/boot/impl/NioBossServerSocketImpl.class
cn/kungreat/boot/CpdogMain.class
cn/kungreat/boot/impl/NioWorkServerSocketImpl$WorkRunnable.class
cn/kungreat/boot/tls/TLSSocketLink.class
cn/kungreat/boot/em/ProtocolState.class
cn/kungreat/boot/impl/NioWorkServerSocketImpl.class
cn/kungreat/boot/utils/JdbcUtils.class
cpdog.properties
cn/kungreat/boot/enents/BaseEvents.class
cn/kungreat/boot/impl/WorkThreadGroup.class
cn/kungreat/boot/utils/CutoverBytes.class
cn/kungreat/boot/exp/WebSocketExceptional.class
cn/kungreat/boot/jb/BaseResponse.class
cn/kungreat/boot/utils/Paging.class
jks-password.txt
cn/kungreat/boot/filter/BaseWebSocketFilter.class
cn/kungreat/boot/jb/MsgDescribe.class
cn/kungreat/boot/utils/WebSocketResponse.class
cn/kungreat/boot/FilterInHandler.class
cn/kungreat/boot/jb/QueryResult.class
cn/kungreat/boot/GlobalEventListener$EventBean.class
cn/kungreat/boot/jb/UserDetails.class
cn/kungreat/boot/GlobalEventListener.class
cn/kungreat/boot/handler/WebSocketConvertDataOut.class
cn/kungreat/boot/handler/WebSocketProtocolHandler$WebSocketBean.class
```

```
cn/kungreat/boot/tls/CpDogSSLContext $ 1.class
META - INF/maven/cn.kungreat/cpdog/pom.xml
cn/kungreat/boot/handler/WebSocketConvertData $ ChartsContent.class
cn/kungreat/boot/NioWorkServerSocket.class
cn/kungreat/boot/services/FriendsService.class
META - INF/maven/cn.kungreat/cpdog/pom.properties
```

注意：通过上面的方式可以遍历 JAR 包内的所有资源数据。

8.6　UDP

UDP 没有可靠性和顺序保证。UDP 在数据传输过程中延迟小、效率高，适合于对可靠性要求不高的应用程序。

8.6.1　DatagramSocket 类

DatagramSocket 类表示用于发送和接收数据报包的套接字。

1. 构造器

DatagramSocket 构造器见表 8-3。

表 8-3　DatagramSocket 构造器

构　造　器	描　　述
DatagramSocket()	构造新的对象，默认为无参构造器
DatagramSocket(int port)	构造新的对象，指定绑定的端口
DatagramSocket(int port, InetAddress laddr)	构造新的对象，指定绑定的端口，指定绑定的地址
DatagramSocket(SocketAddress bindaddr)	构造新的对象，指定绑定的套接字地址

2. 多播示例

基于官方文档示例多播接收方，代码如下：

```
//第 8 章/six/UdpOne.java
public class UdpOne {

    public static void main(String[] args) throws Exception {
        try (DatagramSocket socket = new DatagramSocket(null)) {
            socket.setReuseAddress(true);             //在绑定之前设置地址重用
                //绑定本地所有的网络接口 0.0.0.0
            socket.bind(new InetSocketAddress(6789));
            //广播组 228.5.6.7
            InetAddress castAddr = InetAddress.getByName("228.5.6.7");
            //端口号为 0,会让系统在绑定操作中提取临时端口
            InetSocketAddress group = new InetSocketAddress(castAddr, 0);
```

```
            //获取本地的一个网络接口
            NetworkInterface netIf = NetworkInterface.
               getByInetAddress(InetAddress.getLocalHost());
            //加入广播组
            socket.joinGroup(group, netIf);
            byte[] msgBytes = new byte[1024];            //接收数据的缓冲区
           DatagramPacket packet = new DatagramPacket(msgBytes,
                                                          msgBytes.length);
            //等待接收数据
            socket.receive(packet);
            //输出发送数据方的地址信息
            System.out.println(packet.getSocketAddress());
            //输出接收的数据
            System.out.println(new String(packet.getData(),
                                packet.getOffset(), packet.getLength()));
            //退出广播组
            socket.leaveGroup(group, netIf);
        } catch (Exception exception) {
            exception.printStackTrace();
        }
    }
}
```

基于官方文档示例多播发送方，代码如下：

```
//第8章/six/UdpTwo.java
public class UdpTwo {

    public static void main(String[] args) throws Exception {
        try (DatagramSocket sender = new DatagramSocket(
                                         new InetSocketAddress(9999))) {
            //获取本地的一个网络接口
            NetworkInterface outgoingIf = NetworkInterface.
                        getByInetAddress(InetAddress.getLocalHost());
            //设置 IP 多播
            sender.setOption(StandardSocketOptions.IP_MULTICAST_IF,
                                                            outgoingIf);
            //对于 IPv4 套接字，该选项是套接字发送的多点广播数据报上的生存时间
               //具有零 TTL 的数据报不会在网络上传输，但可以在本地传递
            int ttl = 18; //0 到 255 的数字
            sender.setOption(StandardSocketOptions.IP_MULTICAST_TTL, ttl);
            byte[] msgBytes = "HELLO IP_MULTICAST".getBytes();
            //广播组 228.5.6.7
            InetAddress castAddr = InetAddress.getByName("228.5.6.7");
            //接收方的地址信息
            InetSocketAddress dest = new InetSocketAddress(castAddr, 6789);
            //创建数据报
            DatagramPacket hi = new DatagramPacket(msgBytes,
                                                  msgBytes.length, dest);
            //发送数据报
```

```
            sender.send(hi);
        } catch (Exception e) {
            e.printStackTrace();
        }
    }
}
```

先运行接收方程序再运行发送方程序。

接收方程序的执行结果如下:

```
/192.168.80.1:9999
HELLO IP_MULTICAST
```

3. 常用方法

1) bind(SocketAddress addr)

将当前数据报绑定到指定的套接字地址。接收 SocketAddress 入参,作为指定的套接字地址。

2) close()

关闭当前数据报套接字并释放资源。

3) connect(InetAddress address, int port)

将当前数据报连接到指定的远程地址和端口。接收 InetAddress 入参,作为指定的远程地址,接收 int 入参,作为指定的端口。

4) connect(SocketAddress addr)

将当前数据报连接到指定的远程套接字地址。接收 SocketAddress 入参,作为指定的远程套接字地址。

5) disconnect()

断开远程套接字连接。

6) getBroadcast()

返回当前数据报是否启用 SO_BROAST 选项。

7) getInetAddress()

返回当前数据报所连接的远程地址。

8) getLocalAddress()

返回当前数据报所绑定的本地地址。

9) getLocalPort()

返回当前数据报所绑定的本地端口。

10) getPort()

返回当前数据报所连接的远程端口。

11) getOption(SocketOption<T> name)

返回当前数据报指定选项的值。接收 SocketOption 入参,作为指定选项。

12）getReceiveBufferSize()

返回当前数据报接收数据的缓冲区大小，以字节为单位。

13）getSendBufferSize()

返回当前数据报发送数据的缓冲区大小，以字节为单位。

14）getReuseAddress()

返回当前数据报是否启用SO_REUSEADDR选项。

15）getTrafficClass()

返回当前数据报IP报头中的流量类型或服务类型。

16）isBound()

检查当前数据报是否绑定了本地地址，返回boolean值。

17）isConnected()

检查当前数据报是否连接了远程地址，返回boolean值。

18）isClosed()

检查当前数据报是否已经关闭，返回boolean值。

19）joinGroup(SocketAddress mcastaddr, NetworkInterface netIf)

使当前数据报加入多点广播组。接收SocketAddress入参，作为套接字地址，接收NetworkInterface入参，作为网络接口。

20）leaveGroup(SocketAddress mcastaddr, NetworkInterface netIf)

使当前数据报退出多点广播组。接收SocketAddress入参，作为套接字地址，接收NetworkInterface入参，作为网络接口。

21）receive(DatagramPacket p)

使当前数据报等待接收数据报包。接收DatagramPacket入参，作为数据报包。

22）send(DatagramPacket p)

使当前数据报发送数据报包。接收DatagramPacket入参，作为数据报包。

23）setBroadcast(boolean on)

设置当前数据报是否启用SO_BROAST选项。接收boolean值，作为选项值。

24）supportedOptions()

返回当前数据报所支持的套接字选项集合。

4. 单播示例

单播接收方，代码如下：

```
//第8章/six/UdpTest.java
public class UdpTest {
    public static void main(String[] args) {
        try (DatagramSocket datagramSocket = new DatagramSocket(20066,
                                        InetAddress.getLocalHost())) {
            System.out.println(datagramSocket.isBound());
            System.out.println(datagramSocket.isConnected());
```

```
            byte[] bts = new byte[1024];
            DatagramPacket datagramPacket = new DatagramPacket
                                                (bts, bts.length);
            //等待接收数据
            datagramSocket.receive(datagramPacket);
            //输出接收的数据
            System.out.println(new String(datagramPacket.getData(),
                datagramPacket.getOffset(), datagramPacket.getLength()));
            //输出发送方的地址
            System.out.println(datagramPacket.getSocketAddress());
            //设置新的数据
            datagramPacket.setData("one server".getBytes());
            //将数据回传给发送方
            datagramSocket.send(datagramPacket);
        } catch (Exception e) {
            e.printStackTrace();
        }
    }
}
```

单播发送方,代码如下:

```
//第 8 章/six/UdpTwoTest.java
public class UdpTwoTest {
    static final byte[] BTS = "two server".getBytes();
    static final byte[] BTRFS = new byte[64];

    public static void main(String[] args) {
        try (DatagramSocket datagramSocket = new DatagramSocket(9999,
                                            InetAddress.getLocalHost())) {
            System.out.println(datagramSocket.isBound());
            System.out.println(datagramSocket.isConnected());
            //创建数据报包
            DatagramPacket datagramPacket = new DatagramPacket(BTS,
                        BTS.length, InetAddress.getLocalHost(), 20066);
            //发送数据
            datagramSocket.send(datagramPacket);
            //设置新的数据缓冲区
            datagramPacket.setData(BTRFS);
            //等待接收数据
            datagramSocket.receive(datagramPacket);
            //输出接收的数据
            System.out.println(new String(datagramPacket.getData(),
                datagramPacket.getOffset(), datagramPacket.getLength()));
            //输出发送方的地址
            System.out.println(datagramPacket.getSocketAddress());
        } catch (Exception e) {
            e.printStackTrace();
        }
    }
}
```

先运行接收方程序再运行发送方程序。

接收方程序执行的结果如下：

```
true
false
two server
/192.168.80.1:9999
```

发送方程序执行的结果如下：

```
true
false
one server
/192.168.80.1:20066
```

8.6.2 DatagramPacket 类

DatagramPacket 类表示数据报包配合 DatagramSocket 类使用。

1. 构造器

DatagramPacket 构造器见表 8-4。

表 8-4 DatagramPacket 构造器

构 造 器	描 述
DatagramPacket(byte[] buf, int length)	构造新的对象,指定初始数据,指定数据长度
DatagramPacket(byte[] buf, int offset, int length)	构造新的对象,指定初始数据,指定数据偏移量,指定数据长度
DatagramPacket(byte[] buf, int offset, int length, InetAddress address, int port)	构造新的对象,指定初始数据,指定数据偏移量,指定数据长度,指定绑定的地址,指定绑定的端口
DatagramPacket(byte[] buf, int offset, int length, SocketAddress address)	构造新的对象,指定初始数据,指定数据偏移量,指定数据长度,指定绑定的套接字地址
DatagramPacket(byte[] buf, int length, InetAddress address, int port)	构造新的对象,指定初始数据,指定数据长度,指定绑定的地址,指定绑定的端口
DatagramPacket(byte[] buf, int length, SocketAddress address)	构造新的对象,指定初始数据,指定数据长度,指定绑定的套接字地址

2. 常用方法

1) getAddress()

返回当前数据报包所绑定的地址。

2) getData()

返回当前数据报包的数据缓冲区。

3) getLength()

返回当前数据报包有效的数据长度。

4）getOffset()

返回当前数据报包的数据偏移量。

5）getPort()

返回当前数据报包所绑定的端口。

6）getSocketAddress()

返回当前数据报包所绑定的套接字地址。

7）setAddress(InetAddress iaddr)

设置当前数据报包所绑定的地址。接收 InetAddress 入参，作为地址。

8）setData(byte[] buf)

设置当前数据报包的数据缓冲区。接收 byte[]入参，作为数据缓冲区。

9）setLength(int length)

设置当前数据报包有效的数据长度。接收 int 入参，作为有效的数据长度。

10）setPort(int iport)

设置当前数据报包所绑定的端口。接收 int 入参，作为绑定的端口。

11）setSocketAddress(SocketAddress address)

设置当前数据报包所绑定的套接字地址。接收 SocketAddress 入参，作为套接字地址。

小结

通过本章的学习需要理解网络中的常用概念。

习题

1. 判断题

(1) IP 地址可分为 IPv4 和 IPv6 两种类型。(　　)

(2) 本地的网络接口可能有多个，包括物理网络设备和虚拟网络设备。(　　)

(3) UDP 是一个不稳定的协议，不保证数据的成功发送和到达顺序。(　　)

(4) 0.0.0.0 表示 IPv4 的通配符地址。(　　)

(5) 127.0.0.1 表示 IPv4 的本地回环地址。(　　)

2. 选择题

(1) InetAddress 类获取本地 IP 地址的方法有(　　)？(多选)

A. getByAddress(byte[] addr)　　B. getByName(String host)

C. getAllByName(String host)　　D. isAnyLocalAddress()

(2) InetAddress 类检查当前 IP 地址是否是通配符的方法是(　　)。(单选)

A. isAnyLocalAddress()　　B. isLinkLocalAddress()

C. isLoopbackAddress()　　D. isMulticastAddress()

(3) InetAddress类检查当前IP地址是否是回环地址的方法是(　　)。(单选)

A. isAnyLocalAddress()　　B. isLinkLocalAddress()

C. isLoopbackAddress()　　D. isMulticastAddress()

(4) NetworkInterface类获取本地所有网络接口的方法有(　　)?(多选)

A. getNetworkInterfaces()　　B. networkInterfaces()

C. isUp()　　D. isPointToPoint()

(5) NetworkInterface类检查网络接口是否已启动并正在运行的方法是(　　)。(单选)

A. getNetworkInterfaces()　　B. networkInterfaces()

C. isUp()　　D. isPointToPoint()

(6) 表示统一资源标识符的类是(　　)。(单选)

A. URI类　　B. URL类

C. HttpURLConnection类　　D. JarURLConnection类

3. 填空题

(1) 查看执行结果并补充代码，代码如下：

```
//第8章/answer/OneAnswer.java
public class OneAnswer {

    public static void main(String[] args) throws Exception {
        InetAddress address = InetAddress.________________;
        System.out.println(address);
        System.out.println(address.isLoopbackAddress());
        InetAddress name = InetAddress.__________________;
        System.out.println(name);
        System.out.println(name.isAnyLocalAddress());
    }
}
```

执行结果如下：

```
localhost/127.0.0.1
true
/0.0.0.0
true
```

(2) 查看执行结果并补充代码，代码如下：

```
//第8章/answer/TwoAnswer.java
public class TwoAnswer {

    public static void main(String[] args) throws Throwable {
        URI uri = new URI("______________________________");
        System.out.println(uri.getScheme());
        System.out.println(uri.getSchemeSpecificPart());
```

```
        System.out.println(uri.getAuthority());
        System.out.println(uri.getUserInfo());
        System.out.println(uri.getHost());
        System.out.println(uri.getPort());
        System.out.println(uri.getPath());
        System.out.println(uri.getQuery());
        System.out.println(uri.getFragment());
    }
}
```

执行结果如下：

```
https
//www.example.com:443/languages/java/sample/index.html?name=kungreat
www.example.com:443
null
www.example.com
443
/languages/java/sample/index.html
name=kungreat
28
```

（3）查看执行结果并补充代码，代码如下：

```
//第8章/answer/ThreeAnswer.java
public class ThreeAnswer {
    static final byte[] BTS = "________".getBytes();

    public static void main(String[] args) throws Exception {
        new Thread(_________________________).start();
        Thread.sleep(2000);
        new Thread(ThreeAnswer::callTwo).start();
    }

    private static void callOne() {
        try (DatagramSocket datagramSocket = new DatagramSocket(20066,
                                InetAddress.____________________)) {
            System.out.println(datagramSocket.isBound());
            System.out.println(datagramSocket.isConnected());
            byte[] bts = new byte[1024];
            DatagramPacket datagramPacket = new DatagramPacket(bts,
                                                  bts.length);
            //等待接收数据
            datagramSocket.receive(datagramPacket);
            //输出接收的数据
            System.out.println(new String(datagramPacket.getData(),
                datagramPacket.getOffset(), datagramPacket.getLength()));
            //输出发送方的地址
            System.out.println(datagramPacket.getSocketAddress());
        } catch (Exception e) {
```

```
            e.printStackTrace();
        }
    }

    private static void callTwo() {
        try (DatagramSocket datagramSocket = new DatagramSocket(9999,
                            InetAddress.getLoopbackAddress())) {
            System.out.println(datagramSocket.isBound());
            System.out.println(datagramSocket.isConnected());
            //创建数据报包
            DatagramPacket datagramPacket = new DatagramPacket(BTS,
              BTS.length, InetAddress.____________________, 20066);
            //发送数据
            datagramSocket.send(datagramPacket);
        } catch (Exception e) {
            e.printStackTrace();
        }
    }
}
```

执行结果如下：

```
true
false
true
false
two server
/127.0.0.1:9999
```

第 9 章

CHAPTER 9

Socket 套接字

Socket 套接字是基于 TCP/IP 协议的套接字，套接字是网络之间通信的端点，应用程序可以通过它发送和接收数据。

9.1 ServerSocket 类

此类表示服务器端套接字，服务器端套接字等待客户端套接字请求并通过网络连接，当网络连接完成后便可以进行双向通信。

9.1.1 构造器

ServerSocket 构造器见表 9-1。

表 9-1 ServerSocket 构造器

构　造　器	描　　述
ServerSocket()	构造新的对象，默认为无参构造器
ServerSocket(int port)	构造新的对象，指定绑定的端口
ServerSocket(int port, int backlog)	构造新的对象，指定绑定的端口，指定连接的最大队列长度
ServerSocket(int port, int backlog, InetAddress bindAddr)	构造新的对象，指定绑定的端口，指定连接的最大队列长度，指定绑定的地址

9.1.2 常用方法

本节基于官方文档介绍常用方法，核心方法配合代码示例以方便读者理解。

1. accept()

监听当前绑定的地址和端口，并等待客户端套接字的连接，然后返回表示此连接的套接字对象，代码如下：

```java
//第 9 章/one/ServerSocketTest.java
public class ServerSocketTest {

    public static void main(String[] args) throws InterruptedException {
        new Thread(ServerSocketTest::server).start();
        Thread.sleep(500);
        new Thread(ServerSocketTest::client).start();
    }

    //服务器端
    private static void server() {
        try (ServerSocket serverSocket = new ServerSocket(20066)) {
            System.out.println("服务器端:" + serverSocket.isBound());
            System.out.println("服务器端:" + serverSocket.getInetAddress());
            //监听当前绑定的地址和端口
            Socket accept = serverSocket.accept();
            //输出服务器端的地址信息
            System.out.println("服务器端的地址:" +
                                        accept.getLocalSocketAddress());
            //输出客户端的地址信息
            System.out.println("客户端的地址:" +
                                        accept.getRemoteSocketAddress());
        } catch (Exception e) {
            e.printStackTrace();
        }
    }

    //客户端
    private static void client() {
        try (Socket socket = new Socket()) {
            System.out.println("客户端:" + socket.isBound());
            System.out.println("客户端:" + socket.isConnected());
            //绑定到本地的地址
            socket.bind(new InetSocketAddress(
                            InetAddress.getLoopbackAddress(), 10066));
            //连接到服务器端的地址
            socket.connect(new InetSocketAddress(
                            InetAddress.getLoopbackAddress(), 20066));
            System.out.println("客户端:" + socket.isBound());
            System.out.println("客户端:" + socket.isConnected());
        } catch (Exception e) {
            e.printStackTrace();
        }
    }
}
```

执行结果如下：

```
服务器端:true
服务器端:0.0.0.0/0.0.0.0
```

```
客户端:false
客户端:false
客户端:true
客户端:true
服务器端的地址:/127.0.0.1:20066
客户端的地址:/127.0.0.1:10066
```

2. bind(SocketAddress endpoint)

将当前套接字绑定到指定的套接字地址。接收 SocketAddress 入参,作为指定的套接字地址。

3. bind(SocketAddress endpoint, int backlog)

将当前套接字绑定到指定的套接字地址。接收 SocketAddress 入参,作为指定的套接字地址,接收 int 入参,作为指定连接的最大队列长度。

4. close()

关闭当前套接字并释放资源。

5. getInetAddress()

返回当前套接字绑定的本地地址。

6. getLocalPort()

返回当前套接字绑定的端口。

7. getLocalSocketAddress()

返回当前套接字绑定的本地套接字地址。

8. getOption(SocketOption < T > name)

返回当前套接字指定选项的值。接收 SocketOption 入参,作为指定选项。

9. getReceiveBufferSize()

返回当前套接字接收数据的缓冲区大小,以字节为单位。

10. getReuseAddress()

返回当前套接字是否启用 SO_REUSEADDR 选项。

11. getSoTimeout()

返回当前套接字等待客户端套接字连接的最长超时时间。

12. isBound()

检查当前套接字是否绑定了本地地址,返回 boolean 值。

13. isClosed()

检查当前套接字是否已经关闭,返回 boolean 值。

14. setOption(SocketOption<T> name, T value)

设置当前套接字指定选项的值。接收 SocketOption 入参,作为指定选项,接收 T 入参,作为指定选项的值。

15. setReceiveBufferSize(int size)

设置当前套接字接收数据的缓冲区大小,以字节为单位。接收 int 入参,作为缓冲区长度。

16. setReuseAddress(boolean on)

设置当前套接字是否启用 SO_REUSEADDR 选项。接收 boolean 入参,作为是否启用。

17. setSoTimeout(int timeout)

设置当前套接字等待客户端套接字连接的最长超时时间。接收 int 入参,作为最长超时时间。

注意:当最长超时时间为 0 时表示不限时间。

---

18. supportedOptions()

返回当前套接字所支持的套接字选项集合,代码如下:

```
//第 9 章/one/ServerSocketTest.java
public class ServerSocketTest {

    public static void main(String[] args) throws InterruptedException {
        try (ServerSocket serverSocket = new ServerSocket(20066)) {
            System.out.println(serverSocket.supportedOptions());
        } catch (Exception e) {
            e.printStackTrace();
        }
    }
}
```

执行结果如下:

```
[SO_REUSEADDR, IP_TOS, SO_RCVBUF]
```

9.2 Socket 类

此类表示客户端套接字,通过请求服务器端套接字建立网络连接。当网络连接完成后便可以进行双向通信。

9.2.1 构造器

Socket 构造器见表 9-2。

表 9-2 Socket 构造器

构　造　器	描　　述
Socket()	构造新的对象,默认为无参构造器
Socket(String host, int port)	构造新的对象,指定连接的地址,指定连接的端口
Socket(String host, int port, InetAddress localAddr, int localPort)	构造新的对象,指定连接的地址,指定连接的端口,指定绑定的地址,指定绑定的端口
Socket(InetAddress address, int port)	构造新的对象,指定连接的地址,指定连接的端口
Socket(InetAddress address, int port, InetAddress localAddr, int localPort)	构造新的对象,指定连接的地址,指定连接的端口,指定绑定的地址,指定绑定的端口
Socket(Proxy proxy)	构造新的对象,指定使用的代理

9.2.2 常用方法

本节基于官方文档介绍常用方法,核心方法配合代码示例以方便读者理解。

1. bind(SocketAddress bindpoint)

将当前套接字绑定到指定的套接字地址。接收 SocketAddress 入参,作为指定的套接字地址。

2. close()

关闭当前套接字并释放资源。

3. connect(SocketAddress endpoint)

将当前套接字连接到指定的套接字地址。接收 SocketAddress 入参,作为指定的套接字地址。

4. connect(SocketAddress endpoint, int timeout)

将当前套接字连接到指定的套接字地址。接收 SocketAddress 入参,作为指定的套接字地址,接收 int 入参,作为最长超时时间。

5. getInetAddress()

返回当前套接字所连接的地址。

6. getInputStream()

返回当前套接字所连接的字节输入流,代码如下:

```
//第 9 章/two/SocketTest.java
public class SocketTest {

    public static void main(String[] args) throws Exception {
```

```
        new Thread(SocketTest::serverSocket).start();
        Thread.sleep(500);
        new Thread(SocketTest::clientSocket).start();
    }

    public static void clientSocket() {
        try (Socket socket = new Socket(InetAddress.getLoopbackAddress(),
                20066, InetAddress.getLoopbackAddress(), 10066)) {
            if (socket.isConnected()) {
                System.out.println("client start");
                InputStream inputStream = socket.getInputStream();
                byte[] bytes = inputStream.readAllBytes();
                System.out.println(new String(bytes));
            }
        } catch (Exception e) {
            e.printStackTrace();
        }
    }


    public static void serverSocket() {
        try (ServerSocket serverSocket = new ServerSocket(20066);
             Socket accept = serverSocket.accept()) {
            OutputStream outputStream = accept.getOutputStream();
            outputStream.write(("已经连接到服务器,你的 IP 地址是:" +
        accept.getRemoteSocketAddress().toString()).getBytes());
        } catch (Exception e) {
            e.printStackTrace();
        }
    }
}
```

执行结果如下：

```
client start
已经连接到服务器,你的 IP 地址是:/127.0.0.1:10066
```

注意：套接字所连接的字节输入流读取数据的操作是阻塞式的。

7．getKeepAlive()

返回当前套接字是否启用 SO_KEEPALIVE 选项。

8．getLocalAddress()

返回当前套接字所绑定的地址。

9．getLocalPort()

返回当前套接字所绑定的端口。

10. getLocalSocketAddress()

返回当前套接字所绑定的套接字地址。

11. getOOBInline()

返回当前套接字是否启用 SO_OOBINLINE 选项。

12. getOption(SocketOption < T > name)

返回当前套接字指定选项的值。接收 SocketOption 入参,作为指定选项。

13. getOutputStream()

返回当前套接字所连接的字节输出流。

14. getPort()

返回当前套接字所连接的端口。

15. getReceiveBufferSize()

返回当前套接字接收数据的缓冲区大小,以字节为单位。

16. getRemoteSocketAddress()

返回当前套接字所连接的套接字地址。

17. getReuseAddress()

返回当前套接字是否启用 SO_REUSEADDR 选项。

18. getSendBufferSize()

返回当前套接字发送数据的缓冲区大小,以字节为单位。

19. getSoLinger()

返回当前套接字 SO_LINGER 选项的值。

20. getSoTimeout()

返回当前套接字 SO_TIMEOUT 选项的值。当此值为 0 时表示无超时时间。

21. getTcpNoDelay()

返回当前套接字是否启用 TCP_NODELAY 选项。

22. getTrafficClass()

返回当前套接字 IP 报头中的流量类型或服务类型。

23. isBound()

检查当前套接字是否绑定了本地地址,返回 boolean 值。

24. isClosed()

检查当前套接字是否已经关闭,返回 boolean 值。

25. isConnected()

检查当前套接字是否已经连接,返回 boolean 值。

26. isInputShutdown()

检查当前套接字是否已经关闭了输入流,返回 boolean 值。

27. isOutputShutdown()

检查当前套接字是否已经关闭了输出流,返回 boolean 值。

28. setKeepAlive(boolean on)

设置当前套接字 SO_KEEPALIVE 选项的值。接收 boolean 入参,作为选项的值。

29. setOOBInline(boolean on)

设置当前套接字 SO_OOBINLINE 选项的值。接收 boolean 入参,作为选项的值。

30. setOption(SocketOption <T> name, T value)

设置当前套接字指定选项的值。接收 SocketOption 入参,作为指定选项,接收 T 入参,作为指定选项的值。

31. setReceiveBufferSize(int size)

设置当前套接字接收数据的缓冲区大小,以字节为单位。接收 int 入参,作为缓冲区长度。

32. setReuseAddress(boolean on)

设置当前套接字是否启用 SO_REUSEADDR 选项。接收 boolean 入参,作为是否启用。

33. setSendBufferSize(int size)

设置当前套接字发送数据的缓冲区大小,以字节为单位。接收 int 入参,作为缓冲区长度。

34. setSoLinger(boolean on, int linger)

设置当前套接字是否启用 SO_LINGER 选项。接收 boolean 入参,作为是否启用,接收 int 入参,作为选项的值。

35. setSoTimeout(int timeout)

设置当前套接字 SO_TIMEOUT 选项的值。当此值为 0 时表示无超时时间。接收 int 入参,作为选项的值,代码如下:

```
//第 9 章/two/SocketTest.java
public class SocketTest {

    public static void main(String[] args) throws Exception {
        new Thread(SocketTest::serverSocket).start();
```

```
        Thread.sleep(500);
        new Thread(SocketTest::clientSocket).start();
    }

    public static void clientSocket() {
        try (Socket socket = new Socket(InetAddress.getLoopbackAddress(),
                20066, InetAddress.getLoopbackAddress(), 10066)) {
            socket.setSoTimeout(3000);
            if (socket.isConnected()) {
                System.out.println("client start");
                InputStream inputStream = socket.getInputStream();
                //SO_TIMEOUT 会影响字节输入流的数据读取，如果超过最长超时时间，则抛出异常
                byte[] bytes = inputStream.readAllBytes();
                System.out.println(new String(bytes));
            }
        } catch (Exception e) {
            e.printStackTrace();
        }
    }


    public static void serverSocket() {
        try (ServerSocket serverSocket = new ServerSocket(20066)) {
            //如果没有正确的，则关闭此连接
            Socket accept = serverSocket.accept();
            System.out.println("连接成功:" +
                                        accept.getRemoteSocketAddress());
        } catch (Exception e) {
            e.printStackTrace();
        }
    }
}
```

执行结果如下：

```
client start
连接成功:/127.0.0.1:10066
java.net.SocketTimeoutException: Read timed out
    at java.base/sun.nio.ch.NioSocketImpl.timedRead(NioSocketImpl.java:283)
    at java.base/sun.nio.ch.NioSocketImpl.implRead(NioSocketImpl.java:309)
    at java.base/sun.nio.ch.NioSocketImpl.read(NioSocketImpl.java:350)
    at java.base/sun.nio.ch.NioSocketImpl$1.read(NioSocketImpl.java:803)
    at java.base/java.net.Socket$SocketInputStream.read(Socket.java:966)
    at java.base/java.io.InputStream.readNBytes(InputStream.java:409)
    at java.base/java.io.InputStream.readAllBytes(InputStream.java:346)
    at cn.kungreat.niobook.nine.two.SocketTest.clientSocket(SocketTest.java:25)
    at java.base/java.lang.Thread.run(Thread.java:833)
```

36. setTcpNoDelay(boolean on)

设置当前套接字是否启用 TCP_NODELAY 选项。接收 boolean 入参，作为是否启用。

37. setTrafficClass(int tc)

设置当前套接字 IP 报头中的流量类型或服务类型。接收 int 入参,作为服务类型。

38. shutdownInput()

关闭当前套接字的输入流。

39. shutdownOutput()

关闭当前套接字的输出流。

40. supportedOptions()

返回当前套接字所支持的套接字选项集合,代码如下:

```java
//第 9 章/two/SocketTest.java
public class SocketTest {

    public static void main(String[] args) throws Exception {
        new Thread(SocketTest::serverSocket).start();
        Thread.sleep(500);
        new Thread(SocketTest::clientSocket).start();
    }

    public static void clientSocket() {
        try (Socket socket = new Socket(InetAddress.getLoopbackAddress(),
                20066, InetAddress.getLoopbackAddress(), 10066)) {
            System.out.println(socket.supportedOptions());
        } catch (Exception e) {
            e.printStackTrace();
        }
    }


    public static void serverSocket() {
        try (ServerSocket serverSocket = new ServerSocket(20066)) {
            //如果没有正确的,则关闭此连接
            Socket accept = serverSocket.accept();
            System.out.println("连接成功:" + accept.
                                                getRemoteSocketAddress());
        } catch (Exception e) {
            e.printStackTrace();
        }
    }
}
```

执行结果如下:

```
[TCP_NODELAY, SO_REUSEADDR, SO_LINGER, SO_RCVBUF, SO_KEEPALIVE, SO_SNDBUF, IP_TOS]
连接成功:/127.0.0.1:10066
```

9.2.3 TCP/IP

使用套接字演示 TCP/IP 的流程,代码如下:

```
//第 9 章/two/SocketTest.java
public class SocketTest {

    public static void main(String[] args) throws Exception {
        new Thread(SocketTest::serverSocket).start();
        Thread.sleep(500);
        new Thread(SocketTest::clientSocket).start();
    }

    public static void clientSocket() {
        try (Socket socket = new Socket(InetAddress.getLoopbackAddress(),
                20066, InetAddress.getLoopbackAddress(), 10066)) {
            if (socket.isConnected()) {
                System.out.println("client start");
                InputStream inputStream = socket.getInputStream();
                byte[] bytes = inputStream.readAllBytes();
                System.out.println(new String(bytes));
            }
        } catch (Exception e) {
            e.printStackTrace();
        }
    }


    public static void serverSocket() {
        try (ServerSocket serverSocket = new ServerSocket(20066)) {
            //监听当前绑定的地址和端口
            Socket accept = serverSocket.accept();
            OutputStream outputStream = accept.getOutputStream();
            outputStream.write(("已经连接到服务器,你的 IP 地址是:" +
        accept.getRemoteSocketAddress().toString()).getBytes());
            accept.close();
        } catch (Exception e) {
            e.printStackTrace();
        }
    }
}
```

执行结果如下:

```
client start
已经连接到服务器,你的 IP 地址是:/127.0.0.1:10066
```

使用网络监控软件查看 TCP/IP 协议流程,如图 9-1 所示。

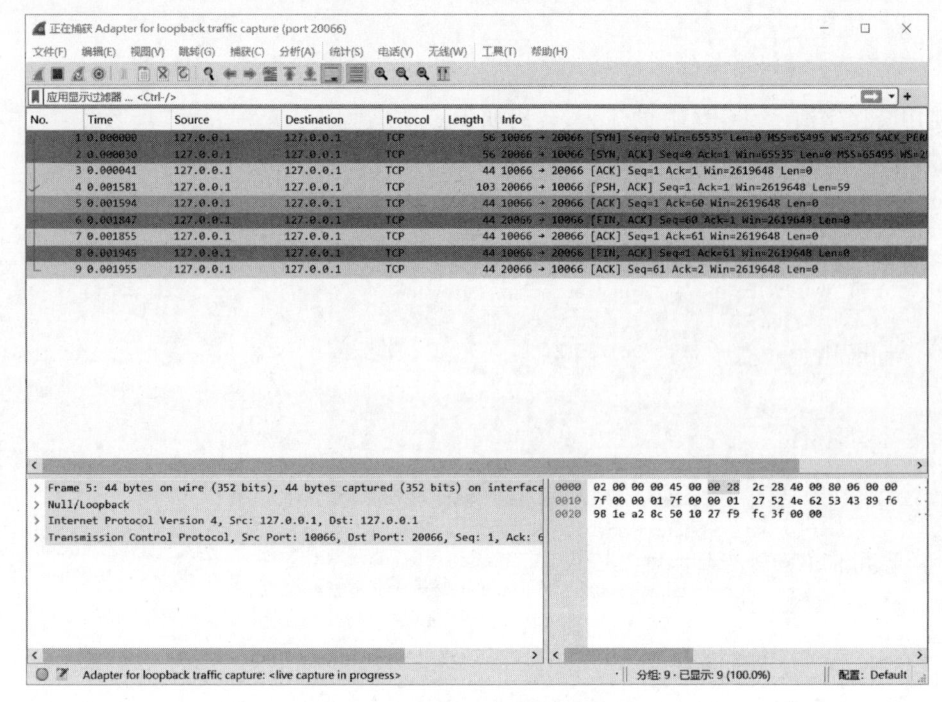

图 9-1 查看 TCP/IP 协议流程

小结

本章需要理解 TCP/IP 协议的概念及网络套接字的使用。

习题

1. 判断题

(1) TCP/IP 协议具有 3 次握手和 4 次挥手的效果。(　　)

(2) UDP 协议也具有 3 次握手和 4 次挥手的效果。(　　)

(3) HTTP1.1 协议依赖于 TCP/IP 协议。(　　)

(4) 端口的使用范围为 0～65 535。(　　)

(5) HTTP 协议的默认端口是 80 端口。(　　)

(6) HTTPS 协议的默认端口是 443 端口。(　　)

2. 选择题

(1) 表示服务器端 TCP/IP 协议套接字的类是(　　)。(单选)

A. Socket 类　　　　B. ServerSocket 类

C. DatagramSocket 类　　D. DatagramPacket 类

(2) 表示 UDP 协议套接字的类是(　　)。(单选)

A. Socket 类　　B. ServerSocket 类

C. DatagramSocket 类　　D. DatagramPacket 类

(3) ServerSocket 类等待客户端连接的方法是(　　)。(单选)

A. accept()　　B. bind(SocketAddress endpoint)

C. getReuseAddress()　　D. isBound()

(4) Socket 类连接服务器端的方法是(　　)。(单选)

A. connect(SocketAddress endpoint)　　B. getInetAddress()

C. getLocalAddress()　　D. isConnected()

(5) Socket 类获取套接字输入流的方法是(　　)。(单选)

A. isInputShutdown()　　B. isOutputShutdown()

C. getInputStream()　　D. getOutputStream()

(6) Socket 类获取套接字输出流的方法是(　　)。(单选)

A. isInputShutdown()　　B. isOutputShutdown()

C. getInputStream()　　D. getOutputStream()

3. 填空题

查看执行结果并补充代码,代码如下:

```
//第 9 章/answer/SocketOne.java
public class SocketOne {

    public static void main(String[] args) throws Exception {
        new Thread(SocketOne::serverSocket).start();
        Thread.sleep(500);
        new Thread(SocketOne::clientSocket).start();
    }

    public static void clientSocket() {
        try (Socket socket = new Socket(InetAddress.getLoopbackAddress(),
                20066, InetAddress.getLoopbackAddress(), 10066)) {
            if (socket.isConnected()) {
                System.out.println("client start");
                InputStream inputStream = socket.getInputStream();
                byte[] bytes = inputStream.________;
                System.out.println(new String(bytes));
            }
        } catch (Exception e) {
            e.printStackTrace();
        }
    }
```

```
    public static void serverSocket() {
        try (ServerSocket serverSocket = new ServerSocket(20066)) {
            //监听当前绑定的地址和端口
            Socket accept = serverSocket.accept();
            OutputStream outputStream = accept.getOutputStream();
            outputStream.write(("已经连接到服务器,你的 IP 地址是:" +
        accept.getRemoteSocketAddress().toString()).getBytes());
            accept.____________________;
        } catch (Exception e) {
            e.printStackTrace();
        }
    }
}
```

执行结果如下：

```
client start
已经连接到服务器,你的 IP 地址是:/127.0.0.1:10066
```

第 10 章

CHAPTER 10

NIO 包

NIO 包定义数据缓冲区的容器、定义字符集及其关联的编码器和解码器、定义各种类型的通道表示与能够执行 IO 操作的实体的连接、定义选择器和选择键与可选通道一起定义复用的非阻塞式 IO 设施。

10.1 Buffer 抽象类

Buffer 抽象类是所有缓冲区的超类，定义了缓冲区的基本功能。核心子类见表 10-1、核心字段见表 10-2。

表 10-1 Buffer 核心子类

类　　名	描　　述
ByteBuffer	字节缓冲区
CharBuffer	字符缓冲区
DoubleBuffer	双精度浮点数缓冲区
FloatBuffer	单精度浮点数缓冲区
IntBuffer	整数缓冲区
LongBuffer	长整数缓冲区
ShortBuffer	短整数缓冲区

表 10-2 Butter 核心字段

字　　段	描　　述
private int capacity	缓冲区的容量
private int limit	数据指针的位置限制
private int position = 0	数据指针的位置
private int mark = −1	数据指针的标记位

本节基于官方文档介绍常用方法，核心方法配合代码示例以方便读者理解。

1. array()

返回当前缓冲区内底层封装的数组。

2．arrayOffset()

返回当前缓冲区内底层封装数组的偏移量。

3．capacity()

返回当前缓冲区内底层封装数组的容量。

4．clear()

重置当前缓冲区内的数据指针。

5．duplicate()

返回当前缓冲区对象的浅复制。

6．flip()

将当前缓冲区的模式切换为输出模式。

7．hasArray()

检查当前缓冲区是否由可访问的数组支持，返回 boolean 值。

8．hasRemaining()

检查当前缓冲区位置与限制之间是否存在可用空间，返回 boolean 值。

9．isDirect()

检查当前缓冲区是否为直接缓冲区，返回 boolean 值。

10．isReadOnly()

检查当前缓冲区是否为只读缓冲区，返回 boolean 值。

11．limit()

返回当前缓冲区数据指针的位置限制。

12．limit(int newLimit)

设置当前缓冲区数据指针的位置限制。接收 int 入参，作为数据指针的位置限制。

13．mark()

设置当前缓冲区的数据指针标记。

14．position()

返回当前缓冲区数据指针的位置。

15．position(int newPosition)

设置当前缓冲区数据指针的位置。接收 int 入参，作为数据指针的位置。

16．remaining()

返回当前缓冲区位置与限制之间的元素数。

17. reset()

将当前缓冲区的数据指针重置为先前标记的值。

18. rewind()

将当前缓冲区的数据指针置为 0 并废弃标记值。

19. slice()

返回当前缓冲区对象的子序列。

20. slice(int index, int length)

返回当前缓冲区对象的子序列。接收 int 入参,作为数据起始索引,接收 int 入参,作为数据长度。

10.2 ByteBuffer 抽象类

56min

字节缓冲区,核心字段见表 10-3。

表 10-3 核心字段

字　段	描　述
final byte[] hb	字节数组缓冲区
final int offset	数组索引的偏移量
boolean isReadOnly	是否为只读缓冲区

本节基于官方文档介绍常用方法,核心方法配合代码示例以方便读者理解。

1. allocate(int capacity)

静态方法,创建指定容量的字节缓冲区对象。接收 int 入参,作为缓冲区的容量,代码如下:

```
//第 10 章/two/ByteBufferTest.java
public class ByteBufferTest {

    public static void main(String[] args) {
        ByteBuffer buffer = ByteBuffer.allocate(1024);
        System.out.println(buffer);
    }
}
```

执行结果如下:

```
java.nio.HeapByteBuffer[pos = 0 lim = 1024 cap = 1024]
```

2. allocateDirect(int capacity)

静态方法,创建指定容量的直接字节缓冲区对象。接收 int 入参,作为缓冲区的容量,代码如下：

```
public class ByteBufferTest {

    public static void main(String[] args) {
        ByteBuffer buffer = ByteBuffer.allocateDirect(1024);
        System.out.println(buffer);
    }
}
```

执行结果如下：

```
java.nio.DirectByteBuffer[pos = 0 lim = 1024 cap = 1024]
```

3. wrap(byte[] array)

静态方法，创建指定初始数据的字节缓冲区对象。接收 byte[]入参，作为缓冲区的初始数据，代码如下：

```
//第 10 章/two/ByteBufferTest.java
public class ByteBufferTest {

    public static void main(String[] args) {
        ByteBuffer buffer = ByteBuffer.wrap("hello".getBytes());
        System.out.println(buffer);
    }
}
```

执行结果如下：

```
java.nio.HeapByteBuffer[pos = 0 lim = 5 cap = 5]
```

4. wrap(byte[] array, int offset, int length)

静态方法，创建指定初始数据的字节缓冲区对象。接收 byte[]入参，作为缓冲区的初始数据，接收 int 入参，作为数组的起始索引，接收 int 入参，作为最大的数据长度。

5. array()

返回当前缓冲区内底层封装的字节数组。

6. arrayOffset()

返回当前缓冲区内底层封装字节数组的偏移量。

7. clear()

重置当前缓冲区内的数据指针，源代码如图 10-1 所示。

```
public Buffer clear() {
    position = 0;
    limit = capacity;
    mark = -1;
    return this;
}
```

图 10-1 源代码

8. compact()

压缩当前缓冲区的数据并切换为输入模式，代码如下：

```
//第 10 章/two/ByteBufferTest.java
public class ByteBufferTest {

    public static void main(String[] args) {
        //默认为输入模式
        ByteBuffer buffer = ByteBuffer.allocate(1024);
        System.out.println(buffer);
        buffer.put("hello ByteBuffer".getBytes());
        System.out.println(buffer);
        buffer.put("hello ByteBuffer".getBytes());
        System.out.println(buffer);

        //切换输出模式
        buffer.flip();
        System.out.println(buffer);
        byte[] btsGet = new byte[buffer.remaining() - 1];
        buffer.get(btsGet);
        System.out.println(new String(btsGet));
        System.out.println(buffer);

        //切换输入模式
        buffer.compact();
        System.out.println(buffer);
    }
}
```

执行结果如下：

```
java.nio.HeapByteBuffer[pos=0 lim=1024 cap=1024]
java.nio.HeapByteBuffer[pos=16 lim=1024 cap=1024]
java.nio.HeapByteBuffer[pos=32 lim=1024 cap=1024]
java.nio.HeapByteBuffer[pos=0 lim=32 cap=1024]
hello ByteBufferhello ByteBuffe
java.nio.HeapByteBuffer[pos=31 lim=32 cap=1024]
java.nio.HeapByteBuffer[pos=1 lim=1024 cap=1024]
```

9. compareTo(ByteBuffer that)

比较当前缓冲区和指定的缓冲区数据是否相同，返回 int 值。接收 ByteBuffer 入参，作为指定的缓冲区。

10. duplicate()

返回当前缓冲区对象的浅复制，代码如下：

```
//第 10 章/two/ByteBufferTest.java
public class ByteBufferTest {

    public static void main(String[] args) {
        //默认为输入模式
        ByteBuffer buffer = ByteBuffer.allocate(1024);
```

```
        buffer.put("hello ByteBuffer".getBytes());
        //切换输出模式
        buffer.flip();
        System.out.println(buffer);
        ByteBuffer duplicate = buffer.duplicate();
        System.out.println((char) duplicate.get());
        System.out.println(duplicate);
        System.out.println(buffer.array() == duplicate.array());
    }
}
```

执行结果如下：

```
java.nio.HeapByteBuffer[pos = 0 lim = 16 cap = 1024]
h
java.nio.HeapByteBuffer[pos = 1 lim = 16 cap = 1024]
true
```

11. flip()

将当前缓冲区的模式切换为输出模式。

12. get()

返回当前缓冲区的字节数据。如果没有数据，则抛出异常，代码如下：

```
//第 10 章/two/ByteBufferTest.java
public class ByteBufferTest {

    public static void main(String[] args) {
        //默认为输入模式
        ByteBuffer buffer = ByteBuffer.allocate(1024);
        System.out.println(buffer);
        buffer.put("hi".getBytes());
        System.out.println(buffer);
        buffer.flip();
        System.out.println(buffer);
        System.out.println((char) buffer.get());
        System.out.println((char) buffer.get());
        //注意观察数据指针
        System.out.println(buffer);
        System.out.println((char) buffer.get());
    }
}
```

执行结果如下：

```
java.nio.HeapByteBuffer[pos = 0 lim = 1024 cap = 1024]
java.nio.HeapByteBuffer[pos = 2 lim = 1024 cap = 1024]
java.nio.HeapByteBuffer[pos = 0 lim = 2 cap = 1024]
h
i
```

```
java.nio.HeapByteBuffer[pos = 2 lim = 2 cap = 1024]
Exception in thread "main" java.nio.BufferUnderflowException
    at java.base/java.nio.Buffer.nextGetIndex(Buffer.java:699)
    at java.base/java.nio.HeapByteBuffer.get(HeapByteBuffer.java:165)
    at cn.kungreat.niobook.ten.two.ByteBufferTest.main(ByteBufferTest.java:19)
```

注意：此方法可以配合 hasRemaining()方法或者 remaining()方法一起使用。

配合使用的代码如下：

```
//第 10 章/two/ByteBufferTest.java
public class ByteBufferTest {

    public static void main(String[] args) {
        //默认为输入模式
        ByteBuffer buffer = ByteBuffer.allocate(1024);
        System.out.println(buffer);
        buffer.put("hi".getBytes());
        System.out.println(buffer);
        buffer.flip();
        System.out.println(buffer);
        while (buffer.hasRemaining()) {
            System.out.println((char) buffer.get());
        }
        System.out.println(buffer);
    }
}
```

执行结果如下：

```
java.nio.HeapByteBuffer[pos = 0 lim = 1024 cap = 1024]
java.nio.HeapByteBuffer[pos = 2 lim = 1024 cap = 1024]
java.nio.HeapByteBuffer[pos = 0 lim = 2 cap = 1024]
h
i
java.nio.HeapByteBuffer[pos = 2 lim = 2 cap = 1024]
```

13. get(byte[] dst)

将当前缓冲区的字节数据存储到指定的字节数组中。接收 byte[]入参，作为指定的字节数组，代码如下：

```
//第 10 章/two/ByteBufferTest.java
public class ByteBufferTest {

    public static void main(String[] args) {
        //默认为输入模式
        ByteBuffer buffer = ByteBuffer.allocate(1024);
        System.out.println(buffer);
```

```
        buffer.put("hello world".getBytes());
        System.out.println(buffer);
        buffer.flip();
        if(buffer.hasRemaining()){
            byte[] bts = new byte[buffer.remaining()];
            buffer.get(bts);
            System.out.println(new String(bts));
        }
        System.out.println(buffer);
    }
}
```

执行结果如下：

```
java.nio.HeapByteBuffer[pos = 0 lim = 1024 cap = 1024]
java.nio.HeapByteBuffer[pos = 11 lim = 1024 cap = 1024]
hello world
java.nio.HeapByteBuffer[pos = 11 lim = 11 cap = 1024]
```

14. get(byte[] dst，int offset，int length)

将当前缓冲区的字节数据存储到指定的字节数组中。接收 byte[]入参，作为指定的字节数组，接收 int 入参，作为数组的起始索引，接收 int 入参，作为最大的数据长度。

15. get(int index)

返回当前缓冲区指定索引的字节数据。如果索引超标，则抛出异常。接收 int 入参，作为指定索引，代码如下：

```
//第 10 章/two/ByteBufferTest.java
public class ByteBufferTest {

    public static void main(String[] args) {
        //默认为输入模式
        ByteBuffer buffer = ByteBuffer.allocate(1024);
        System.out.println(buffer);
        buffer.put("hello world".getBytes());
        System.out.println(buffer);
        buffer.flip();
        System.out.println(buffer);
        //此方法不会修改数据指针的位置
        System.out.println(buffer.get(5));
        System.out.println(buffer.get(3));
        System.out.println(buffer);
    }
}
```

执行结果如下：

```
java.nio.HeapByteBuffer[pos = 0 lim = 1024 cap = 1024]
java.nio.HeapByteBuffer[pos = 11 lim = 1024 cap = 1024]
```

```
java.nio.HeapByteBuffer[pos = 0 lim = 11 cap = 1024]
32
108
java.nio.HeapByteBuffer[pos = 0 lim = 11 cap = 1024]
```

16. get(int index，byte[] dst)

将当前缓冲区的字节数据存储到指定的字节数组中。接收 int 入参，作为缓冲区的起始索引，接收 byte[]入参，作为指定的字节数组，代码如下：

```java
//第 10 章/two/ByteBufferTest.java
public class ByteBufferTest {

    public static void main(String[] args) {
        //默认为输入模式
        ByteBuffer buffer = ByteBuffer.allocate(1024);
        System.out.println(buffer);
        buffer.put("hello world".getBytes());
        System.out.println(buffer);
        buffer.flip();
        System.out.println(buffer);
        byte[] bts = new byte[5];
        //此方法不会修改数据指针的位置
        buffer.get(0, bts);
        System.out.println(new String(bts));
        System.out.println(buffer.get(3));
        System.out.println(buffer);
    }
}
```

执行结果如下：

```
java.nio.HeapByteBuffer[pos = 0 lim = 1024 cap = 1024]
java.nio.HeapByteBuffer[pos = 11 lim = 1024 cap = 1024]
java.nio.HeapByteBuffer[pos = 0 lim = 11 cap = 1024]
hello
108
java.nio.HeapByteBuffer[pos = 0 lim = 11 cap = 1024]
```

17. getChar()

返回当前缓冲区的字符数据。如果没有数据，则抛出异常，代码如下：

```java
//第 10 章/two/ByteBufferTest.java
public class ByteBufferTest {

    public static void main(String[] args) {
        //默认为输入模式
        ByteBuffer buffer = ByteBuffer.allocate(1024);
        System.out.println(buffer);
```

```
        buffer.putChar('h').putChar('i');
        System.out.println(buffer);
        buffer.flip();
        //1 个字符占 2 字节
        System.out.println(buffer.getChar());
        System.out.println(buffer.getChar());
        System.out.println(buffer);
    }
}
```

执行结果如下：

```
java.nio.HeapByteBuffer[pos = 0 lim = 1024 cap = 1024]
java.nio.HeapByteBuffer[pos = 4 lim = 1024 cap = 1024]
h
i
java.nio.HeapByteBuffer[pos = 4 lim = 4 cap = 1024]
```

18. getDouble()

返回当前缓冲区的双精度浮点数据。如果没有数据，则抛出异常。

19. getFloat()

返回当前缓冲区的单精度浮点数据。如果没有数据，则抛出异常。

20. getInt()

返回当前缓冲区的整数型数据。如果没有数据，则抛出异常。

21. getLong()

返回当前缓冲区的长整数型数据。如果没有数据，则抛出异常。

22. getShort()

返回当前缓冲区的短整数型数据。如果没有数据，则抛出异常。

23. getChar(int index)

返回当前缓冲区指定索引的字符数据。如果索引超标，则抛出异常。接收 int 入参，作为指定索引，代码如下：

```
//第 10 章/two/ByteBufferTest.java
public class ByteBufferTest {

    public static void main(String[] args) {
        //默认为输入模式
        ByteBuffer buffer = ByteBuffer.allocate(1024);
        System.out.println(buffer);
        buffer.putChar('h').putChar('i');
        System.out.println(buffer);
        buffer.flip();
        //此方法不会修改数据指针的位置
        System.out.println(buffer.getChar(0));
```

```
        System.out.println(buffer.getChar(2));
        System.out.println(buffer.getChar(1));
        System.out.println(buffer);
    }
}
```

执行结果如下：

```
java.nio.HeapByteBuffer[pos = 0 lim = 1024 cap = 1024]
java.nio.HeapByteBuffer[pos = 4 lim = 1024 cap = 1024]
h
i
栀
java.nio.HeapByteBuffer[pos = 0 lim = 4 cap = 1024]
```

注意：使用此方法需要关注入参的索引值。

---

24. getDouble(int index)

返回当前缓冲区指定索引的双精度浮点数据。接收 int 入参，作为指定索引。如果索引超标，则抛出异常。

25. getFloat(int index)

返回当前缓冲区指定索引的单精度浮点数据。接收 int 入参，作为指定索引。如果索引超标，则抛出异常。

26. getInt(int index)

返回当前缓冲区指定索引的整数型数据。接收 int 入参，作为指定索引。如果索引超标，则抛出异常。

27. getLong(int index)

返回当前缓冲区指定索引的长整数型数据。接收 int 入参，作为指定索引。如果索引超标，则抛出异常。

28. getShort(int index)

返回当前缓冲区指定索引的短整数型数据。接收 int 入参，作为指定索引。如果索引超标，则抛出异常。

29. hasArray()

检查当前缓冲区是否由可访问的字节数组支持，返回 boolean 值。

30. isDirect()

检查当前缓冲区是否为直接缓冲区，返回 boolean 值。

31. limit(int newLimit)

设置当前缓冲区数据指针的位置限制。接收 int 入参，作为数据指针的位置限制。

32. mark()

设置当前缓冲区的数据指针标记。

33. order()

返回当前缓冲区的排序规则。

34. order(ByteOrder bo)

设置当前缓冲区的排序规则。接收 ByteOrder 入参，作为排序规则。

35. position(int newPosition)

设置当前缓冲区数据指针的位置。接收 int 入参，作为数据指针的位置。

36. position()

返回当前缓冲区数据指针的位置。

37. put(byte b)

添加当前缓冲区的字节数据。接收 byte 入参，作为字节数据。

38. put(byte[] src)

添加当前缓冲区的字节数组数据。接收 byte[]入参，作为字节数组数据。

39. put(byte[] src, int offset, int length)

添加当前缓冲区的字节数组数据。接收 byte[]入参，作为字节数组数据，接收 int 入参，作为数组的起始索引，接收 int 入参，作为最大的数据长度。

40. put(int index, byte b)

修改当前缓冲区指定索引的字节数据。接收 int 入参，作为指定索引，接收 byte 入参，作为字节数据，代码如下：

```java
//第 10 章/two/ByteBufferTest.java
public class ByteBufferTest {

    public static void main(String[] args) {
        //默认为输入模式
        ByteBuffer buffer = ByteBuffer.allocate(1024);
        buffer.put("hello world".getBytes());
        System.out.println(buffer);
        buffer.put(6, (byte) 89);
        System.out.println(buffer);
        buffer.flip();
        byte[] bts = new byte[buffer.remaining()];
        buffer.get(bts);
        System.out.println(new String(bts));
    }
}
```

执行结果如下：

```
java.nio.HeapByteBuffer[pos = 11 lim = 1024 cap = 1024]
java.nio.HeapByteBuffer[pos = 11 lim = 1024 cap = 1024]
hello Yorld
```

41. put(int index, byte[] src)

修改当前缓冲区指定索引的字节数组数据。接收 int 入参,作为指定索引,接收 byte[] 入参,作为字节数组数据。

42. put(ByteBuffer src)

添加当前缓冲区的数据。接收 ByteBuffer 入参,作为缓冲区的数据,代码如下:

```java
//第 10 章/two/ByteBufferTest.java
public class ByteBufferTest {

    public static void main(String[] args) {
        //默认为输入模式
        ByteBuffer buffer = ByteBuffer.allocate(1024);
        buffer.put("hello world".getBytes());
        System.out.println(buffer);
        buffer.flip();
        ByteBuffer allocate = ByteBuffer.allocate(64);
        allocate.put(buffer);
        System.out.println(allocate);
    }
}
```

执行结果如下:

```
java.nio.HeapByteBuffer[pos = 11 lim = 1024 cap = 1024]
java.nio.HeapByteBuffer[pos = 11 lim = 64 cap = 64]
```

43. putChar(char value)

添加当前缓冲区的字符数据。接收 char 入参,作为字符数据。

44. putDouble(double value)

添加当前缓冲区的双精度浮点数据。接收 double 入参,作为双精度浮点数据。

45. putFloat(float value)

添加当前缓冲区的单精度浮点数据。接收 float 入参,作为单精度浮点数据。

46. putInt(int value)

添加当前缓冲区的整数型数据。接收 int 入参,作为整数型数据。

47. putLong(long value)

添加当前缓冲区的长整数型数据。接收 long 入参,作为长整数型数据。

48. putShort(short value)

添加当前缓冲区的短整数型数据。接收 short 入参,作为短整数型数据。

49. putChar(int index, char value)

修改当前缓冲区指定索引的字符数据。接收 int 入参，作为指定索引，接收 char 入参，作为字符数据。

50. putDouble(int index, double value)

修改当前缓冲区指定索引的双精度浮点数据。接收 int 入参，作为指定索引，接收 double 入参，作为双精度浮点数据。

51. putFloat(int index, float value)

修改当前缓冲区指定索引的单精度浮点数据。接收 int 入参，作为指定索引，接收 float 入参，作为单精度浮点数据。

52. putInt(int index, int value)

修改当前缓冲区指定索引的整数型数据。接收 int 入参，作为指定索引，接收 int 入参，作为整数型数据。

53. putLong(int index, long value)

修改当前缓冲区指定索引的长整数型数据。接收 int 入参，作为指定索引，接收 long 入参，作为长整数型数据。

54. putShort(int index, short value)

修改当前缓冲区指定索引的短整数型数据。接收 int 入参，作为指定索引，接收 short 入参，作为短整数型数据。

55. reset()

将当前缓冲区的数据指针重置为先前标记的值。

56. rewind()

将当前缓冲区的数据指针置为 0 并废弃标记值。

57. slice()

返回当前缓冲区对象的子序列，代码如下：

```
//第 10 章/two/ByteBufferTest.java
public class ByteBufferTest {

    public static void main(String[] args) {
        //默认为输入模式
        ByteBuffer buffer = ByteBuffer.allocate(1024);
        buffer.put("hello world".getBytes());
        buffer.flip();
        System.out.println(buffer);
        ByteBuffer slice = buffer.slice();
        System.out.println(slice);
```

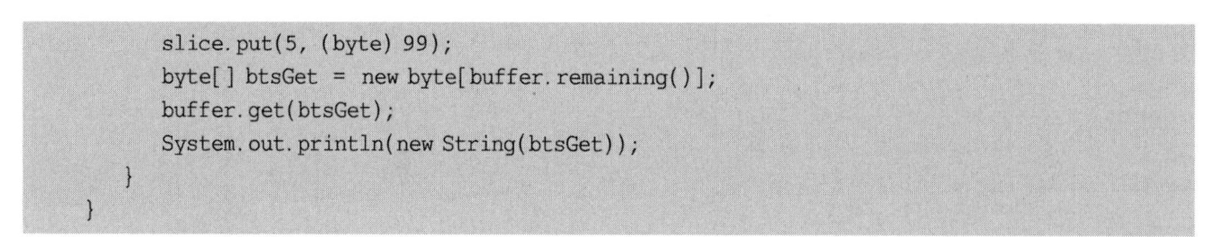

```
        slice.put(5, (byte) 99);
        byte[ ] btsGet = new byte[buffer.remaining()];
        buffer.get(btsGet);
        System.out.println(new String(btsGet));
    }
}
```

执行结果如下：

```
java.nio.HeapByteBuffer[pos = 0 lim = 11 cap = 1024]
java.nio.HeapByteBuffer[pos = 0 lim = 11 cap = 11]
hellocworld
```

58. slice(int index, int length)

返回当前缓冲区对象的子序列。接收 int 入参,作为数据起始索引,接收 int 入参,作为数据长度。

59. hasRemaining()

检查当前缓冲区位置与限制之间是否存在可用空间,返回 boolean 值。

60. remaining()

返回当前缓冲区位置与限制之间的元素数。

10.3 CharBuffer 抽象类

字符缓冲区,核心字段见表 10-4。

表 10-4 核心字段

字 段	描 述
final char[] hb	字符数组缓冲区
final int offset	数组索引的偏移量
boolean isReadOnly	是否为只读缓冲区

本节基于官方文档介绍常用方法,核心方法配合代码示例以方便读者理解。

1. allocate(int capacity)

静态方法,创建指定容量的字符缓冲区对象。接收 int 入参,作为缓冲区的容量,代码如下：

```
//第 10 章/three/CharBufferTest.java
public class CharBufferTest {

    public static void main(String[ ] args) {
        CharBuffer allocate = CharBuffer.allocate(1024);
```

```
        allocate.put('a').put('b').put('c').append("def");
        allocate.flip();
        System.out.println(allocate.position() + " " + allocate.limit());
        System.out.println(allocate);
        System.out.println(allocate.position() + " " + allocate.limit());
    }
}
```

执行结果如下：

```
0 6
abcdef
0 6
```

2．wrap(char[] array)

静态方法，创建指定初始数据的字符缓冲区对象。接收 char[]入参，作为缓冲区的初始数据，代码如下：

```
//第 10 章/three/CharBufferTest.java
public class CharBufferTest {

    public static void main(String[] args) {
        CharBuffer allocate = CharBuffer.wrap(new char[]{'a','b','c','d'});
        System.out.println(allocate.position() + " " + allocate.limit());
        System.out.println(allocate);
        System.out.println(allocate.position() + " " + allocate.limit());
        System.out.println(allocate.limit());
    }
}
```

执行结果如下：

```
0 4
abcd
0 4
4
```

3．wrap(char[] array，int offset，int length)

静态方法，创建指定初始数据的字符缓冲区对象。接收 char[]入参，作为缓冲区的初始数据，接收 int 入参，作为数组的起始索引，接收 int 入参，作为最大的数据长度，代码如下：

```
//第 10 章/three/CharBufferTest.java
public class CharBufferTest {

    public static void main(String[] args) {
        CharBuffer allocate = CharBuffer.wrap(new
                                    char[]{'a','b','c','d','e'},0,4);
```

```
        System.out.println(allocate.position() + " " + allocate.limit());
        System.out.println(allocate);
        System.out.println(allocate.position() + " " + allocate.limit());
        System.out.println(allocate.limit());
    }
}
```

执行结果如下：

```
0 4
abcd
0 4
4
```

4. wrap(CharSequence csq)

静态方法，创建指定初始数据的字符缓冲区对象。接收 CharSequence 入参，作为缓冲区的初始数据，代码如下：

```
//第 10 章/three/CharBufferTest.java
public class CharBufferTest {

    public static void main(String[] args) {
        CharBuffer allocate = CharBuffer.wrap("abcd");
        System.out.println(allocate.position() + " " + allocate.limit());
        System.out.println(allocate);
        System.out.println(allocate.position() + " " + allocate.limit());
        System.out.println(allocate.limit());
    }
}
```

执行结果如下：

```
0 4
abcd
0 4
4
```

5. wrap(CharSequence csq, int start, int end)

静态方法，创建指定初始数据的字符缓冲区对象。接收 CharSequence 入参，作为缓冲区的初始数据，接收 int 入参，作为数据的起始索引，接收 int 入参，作为最大的数据长度，代码如下：

```
//第 10 章/three/CharBufferTest.java
public class CharBufferTest {

    public static void main(String[] args) {
        CharBuffer allocate = CharBuffer.wrap("abcde",0,4);
        System.out.println(allocate.position() + " " + allocate.limit());
```

```
        System.out.println(allocate);
        System.out.println(allocate.position() + " " + allocate.limit());
        System.out.println(allocate.limit());
    }
}
```

执行结果如下：

```
0 4
abcd
0 4
4
```

6. array()

返回当前缓冲区内底层封装的字符数组。

7. arrayOffset()

返回当前缓冲区内底层封装字符数组的偏移量。

8. append(char c)

添加当前缓冲区的字符数据。接收 char 入参，作为字符数据，代码如下：

```
//第 10 章/three/CharBufferTest.java
public class CharBufferTest {

    public static void main(String[] args) {
        CharBuffer allocate = CharBuffer.allocate(1024);
        System.out.println(allocate.position() + " " + allocate.limit()
                                        + " " + allocate.capacity());
        allocate.append('a').append('b').append('c').append('d');
        System.out.println(allocate.position() + " " + allocate.limit()
                                        + " " + allocate.capacity());
        allocate.flip();
        System.out.println(allocate.position() + " " + allocate.limit()
                                        + " " + allocate.capacity());
        System.out.println(allocate);
        System.out.println(allocate.position() + " " + allocate.limit()
                                        + " " + allocate.capacity());
    }
}
```

执行结果如下：

```
0 1024 1024
4 1024 1024
0 4 1024
abcd
0 4 1024
```

9. append(CharSequence csq)

添加当前缓冲区的字符数据。接收 CharSequence 入参,作为字符数据,代码如下:

10. append(CharSequence csq, int start, int end)

添加当前缓冲区的字符数据。接收 CharSequence 入参,作为字符数据,接收 int 入参,作为数据的起始索引,接收 int 入参,作为最大的数据长度。

11. asReadOnlyBuffer()

创建新的只读字符缓冲区并共享当前字符缓冲区的内容,代码如下:

```
//第 10 章/three/CharBufferTest.java
public class CharBufferTest {

    public static void main(String[] args) {
        CharBuffer allocate = CharBuffer.allocate(1024);
        System.out.println(allocate.position() + " " + allocate.limit()
                                         + " " + allocate.capacity());
        allocate.append("abcd");
        System.out.println(allocate.position() + " " + allocate.limit() +
                                        " " + allocate.capacity());
        allocate.flip();
        System.out.println(allocate.position() + " " + allocate.limit() +
                                        " " + allocate.capacity());
        System.out.println(allocate);
        CharBuffer readBuffer = allocate.asReadOnlyBuffer();
        System.out.println(readBuffer);
        System.out.println(readBuffer.position() + " " + readBuffer.limit()
                                         + " " + readBuffer.capacity());
        System.out.println(readBuffer.get());
        System.out.println(readBuffer.position() + " " + readBuffer.limit()
                                         + " " + readBuffer.capacity());
    }
}
```

执行结果如下:

```
0 1024 1024
4 1024 1024
0 4 1024
abcd
abcd
0 4 1024
a
1 4 1024
```

12. charAt(int index)

返回相对于当前位置的指定索引处的字符。接收 int 入参,作为指定索引,代码如下:

```java
//第 10 章/three/CharBufferTest.java
public class CharBufferTest {

    public static void main(String[] args) {
        CharBuffer allocate = CharBuffer.allocate(1024);
        allocate.append("abcd");
        allocate.flip();
        System.out.println(allocate.charAt(1));
        System.out.println(allocate.position() + " " + allocate.limit() +
                                        " " + allocate.capacity());
    }
}
```

执行结果如下：

```
b
0 4 1024
```

13. clear()

重置当前缓冲区内的数据指针。

14. compact()

压缩当前缓冲区的数据并切换为输入模式，代码如下：

```java
//第 10 章/three/CharBufferTest.java
public class CharBufferTest {

    public static void main(String[] args) {
        //默认为输入模式
        CharBuffer buffer = CharBuffer.allocate(1024);
        buffer.append("hello CharBuffer");
        //切换输出模式
        buffer.flip();
        System.out.println(buffer);
        char[] btsGet = new char[buffer.remaining() - 1];
        buffer.get(btsGet);
        System.out.println(new String(btsGet));
        System.out.println(buffer);
        //切换输入模式
        buffer.compact();
        System.out.println(buffer.position() + " " + buffer.limit()
                                    + " " + buffer.capacity());
    }
}
```

执行结果如下：

```
hello CharBuffer
hello CharBuffe
r
1 1024 1024
```

15．compareTo(CharBuffer that)

比较当前缓冲区和指定的缓冲区数据是否相同，返回 int 值。接收 CharBuffer 入参，作为指定的缓冲区。

16．duplicate()

返回当前缓冲区对象的浅复制，代码如下：

```
//第 10 章/three/CharBufferTest.java
public class CharBufferTest {

    public static void main(String[] args) {
        //默认为输入模式
        CharBuffer buffer = CharBuffer.allocate(1024);
        buffer.put("hello ByteBuffer");
        //切换输出模式
        buffer.flip();
        System.out.println(buffer);
        CharBuffer duplicate = buffer.duplicate();
        System.out.println(duplicate.get());
        System.out.println(duplicate);
        System.out.println(buffer.array() == duplicate.array());
    }
}
```

执行结果如下：

```
hello ByteBuffer
h
ello ByteBuffer
true
```

17．flip()

将当前缓冲区的模式切换为输出模式。

18．get()

返回当前缓冲区的字符数据。如果没有数据，则抛出异常，代码如下：

```
//第 10 章/three/CharBufferTest.java
public class CharBufferTest {

    public static void main(String[] args) {
        //默认为输入模式
        CharBuffer buffer = CharBuffer.allocate(1024);
        buffer.put("hello ByteBuffer");
        //切换输出模式
        buffer.flip();
        System.out.println(buffer.get());
        System.out.println(buffer.get());
```

```
        System.out.println(buffer);
    }
}
```

执行结果如下：

```
h
e
llo ByteBuffer
```

19. get(char[] dst)

将当前缓冲区的字符数据存储到指定的字符数组中。接收 char[]入参，作为指定的字符数组，代码如下：

```
//第 10 章/three/CharBufferTest.java
public class CharBufferTest {

    public static void main(String[] args) {
        CharBuffer allocate = CharBuffer.allocate(1024);
        allocate.put("hello world");
        allocate.flip();
        char[] chars = new char[allocate.remaining()];
        allocate.get(chars);
        System.out.println(new String(chars));
        System.out.println(allocate.position() + " " + allocate.limit());
        allocate.compact();
        System.out.println(allocate.position() + " " + allocate.limit());
        System.out.println(allocate.hasArray());
        System.out.println(allocate.isEmpty());
        System.out.println(allocate.length());
    }
}
```

执行结果如下：

```
hello world
11 11
0 1024
true
false
1024
```

20. get(char[] dst, int offset, int length)

将当前缓冲区的字符数据存储到指定的字符数组中。接收 char[]入参，作为指定的字符数组，接收 int 入参，作为数组的起始索引，接收 int 入参，作为最大的数据长度。

21. get(int index)

返回当前缓冲区指定索引的字符数据。如果索引超标，则抛出异常。接收 int 入参，作为指定索引，代码如下：

```
//第 10 章/three/CharBufferTest.java
public class CharBufferTest {

    public static void main(String[] args) {
        CharBuffer allocate = CharBuffer.allocate(1024);
        allocate.put("hello world");
        allocate.flip();
        //此方法不会修改数据指针位置
        System.out.println(allocate.get(1));
        System.out.println(allocate.position() + " " + allocate.limit());
        //此方法会修改数据指针位置
        System.out.println(allocate.get());
        System.out.println(allocate.position() + " " + allocate.limit());
    }
}
```

执行结果如下：

```
e
0 11
h
1 11
```

22. get(int index, char[] dst)

将当前缓冲区的字符数据存储到指定的字符数组中。接收 int 入参，作为缓冲区的起始索引，接收 char[]入参，作为指定的字符数组。

23. hasArray()

检查当前缓冲区是否由可访问的字节数组支持，返回 boolean 值。

24. isDirect()

检查当前缓冲区是否为直接缓冲区，返回 boolean 值。

25. isEmpty()

检查当前缓冲区数据是否为空，返回 boolean 值。

26. limit(int newLimit)

设置当前缓冲区数据指针的位置限制。接收 int 入参，作为数据指针的位置限制。

27. mark()

设置当前缓冲区的数据指针标记。

28. mismatch(CharBuffer that)

返回当前缓冲区与指定缓冲区之间第 1 个不匹配数据的相对索引。接收 CharBuffer 入参，作为指定缓冲区，代码如下：

```java
//第10章/three/CharBufferTest.java
public class CharBufferTest {

    public static void main(String[] args) {
        CharBuffer one = CharBuffer.allocate(1024);
        one.put("hello world");
        CharBuffer two = CharBuffer.allocate(1024);
        two.put("hello world");
        CharBuffer three = CharBuffer.allocate(1024);
        three.put("hello");
        one.flip();
        two.flip();
        three.flip();
        System.out.println(one.mismatch(two));
        System.out.println(one.mismatch(three));
    }
}
```

执行结果如下：

```
-1
5
```

29. order()

返回当前缓冲区的排序规则。

30. position(int newPosition)

设置当前缓冲区数据指针的位置。接收 int 入参，作为数据指针的位置。

31. position()

返回当前缓冲区数据指针的位置。

32. put(char c)

添加当前缓冲区的字符数据。接收 char 入参，作为字符数据。

33. put(char[] src)

添加当前缓冲区的字符数组数据。接收 char[]入参，作为字符数组数据。

34. put(char[] src, int offset, int length)

添加当前缓冲区的字符数组数据。接收 char[]入参，作为字符数组数据，接收 int 入参，作为数组的起始索引，接收 int 入参，作为最大的数据长度。

35. put(int index, char c)

修改当前缓冲区指定索引的字符数据。接收 int 入参，作为指定索引，接收 char 入参，作为字符数据，代码如下：

```java
//第10章/three/CharBufferTest.java
public class CharBufferTest {
```

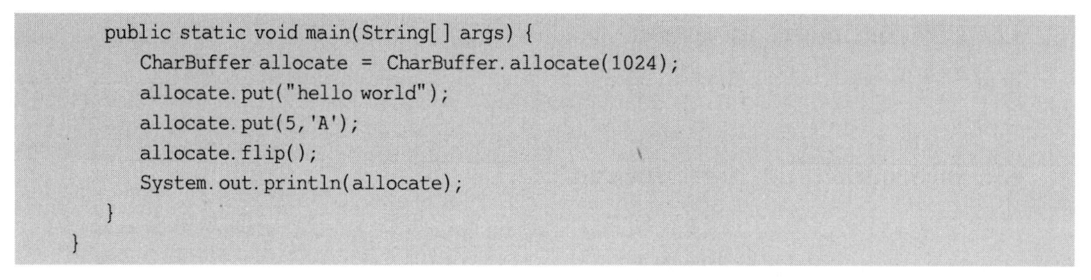

```
    public static void main(String[] args) {
        CharBuffer allocate = CharBuffer.allocate(1024);
        allocate.put("hello world");
        allocate.put(5,'A');
        allocate.flip();
        System.out.println(allocate);
    }
}
```

执行结果如下：

```
helloAworld
```

36. put(int index, char[] src)

修改当前缓冲区指定索引的字符数组数据。接收 int 入参,作为指定索引,接收 char[] 入参,作为字符数组数据。

37. put(String src)

添加当前缓冲区的字符数据。接收 String 入参,作为字符数据。

38. put(String src, int start, int end)

添加当前缓冲区的字符数据。接收 String 入参,作为字符数据,接收 int 入参,作为起始索引,接收 int 入参,作为结束索引,代码如下：

```
//第 10 章/three/CharBufferTest.java
public class CharBufferTest {

    public static void main(String[] args) {
        CharBuffer allocate = CharBuffer.allocate(1024);
        allocate.put("hello world", 0, 5);
        allocate.flip();
        System.out.println(allocate);
    }
}
```

执行结果如下：

```
hello
```

39. reset()

将当前缓冲区的数据指针重置为先前标记的值。

40. rewind()

将当前缓冲区的数据指针置为 0 并废弃标记值。

41. slice()

返回当前缓冲区对象的子序列。

42. slice(int index, int length)

返回当前缓冲区对象的子序列。接收 int 入参，作为数据起始索引，接收 int 入参，作为数据长度。

43. subSequence(int start, int end)

返回当前缓冲区对象的子序列，相对于当前位置。接收 int 入参，作为数据起始索引，接收 int 入参，作为结束索引，代码如下：

```
//第 10 章/three/CharBufferTest.java
public class CharBufferTest {

    public static void main(String[] args) {
        CharBuffer allocate = CharBuffer.allocate(1024);
        allocate.put("hello world");
        allocate.flip();
        System.out.println(allocate.get());
        CharBuffer subBuffer = allocate.subSequence(0,
                                        allocate.remaining());
        System.out.println(subBuffer);
    }
}
```

执行结果如下：

```
h
ello world
```

44. hasRemaining()

检查当前缓冲区位置与限制之间是否存在可用空间，返回 boolean 值。

45. remaining()

返回当前缓冲区位置与限制之间的元素数。

46. isReadOnly()

检查当前缓冲区是否为只读缓冲区，返回 boolean 值。

10.4 IntBuffer 抽象类

整数缓冲区，核心字段见表 10-5。

表 10-5 核心字段

字 段	描 述
final int[] hb	整数数组缓冲区
final int offset	数组索引的偏移量
boolean isReadOnly	是否为只读缓冲区

本节基于官方文档介绍常用方法，核心方法配合代码示例以方便读者理解。

1. allocate(int capacity)

静态方法，创建指定容量的整数缓冲区对象。接收 int 入参，作为缓冲区的容量，代码如下：

```
//第 10 章/four/IntBufferTest.java
public class IntBufferTest {

    public static void main(String[] args) {
        IntBuffer allocate = IntBuffer.allocate(1024);
        System.out.println(allocate);
        allocate.put(56).put(66).put(76);
        System.out.println(allocate);
        allocate.flip();
        System.out.println(allocate);
    }
}
```

执行结果如下：

```
java.nio.HeapIntBuffer[pos = 0 lim = 1024 cap = 1024]
java.nio.HeapIntBuffer[pos = 3 lim = 1024 cap = 1024]
java.nio.HeapIntBuffer[pos = 0 lim = 3 cap = 1024]
```

2. wrap(int[] array)

静态方法，创建指定初始数据的整数缓冲区对象。接收 int[]入参，作为缓冲区的初始数据，代码如下：

```
//第 10 章/four/IntBufferTest.java
public class IntBufferTest {

    public static void main(String[] args) {
        IntBuffer allocate = IntBuffer.wrap(new int[]{55, 65, 75});
        System.out.println(allocate);
        allocate.flip();
        System.out.println(allocate);
        while (allocate.hasRemaining()){
            System.out.println(allocate.get());
        }
        System.out.println(allocate);
    }
}
```

执行结果如下：

```
java.nio.HeapIntBuffer[pos = 0 lim = 3 cap = 3]
java.nio.HeapIntBuffer[pos = 0 lim = 0 cap = 3]
java.nio.HeapIntBuffer[pos = 0 lim = 0 cap = 3]
```

3. wrap(int[] array, int offset, int length)

静态方法,创建指定初始数据的整数缓冲区对象。接收 int[]入参,作为缓冲区的初始数据,接收 int 入参,作为数组的起始索引,接收 int 入参,作为最大的数据长度。

4. array()

返回当前缓冲区内底层封装的整数数组。

5. arrayOffset()

返回当前缓冲区内底层封装整数数组的偏移量。

6. asReadOnlyBuffer()

创建新的只读整数缓冲区并共享当前整数缓冲区的内容,代码如下：

```
//第 10 章/four/IntBufferTest.java
public class IntBufferTest {

    public static void main(String[] args) {
        IntBuffer allocate = IntBuffer.allocate(1024);
        allocate.put(56).put(66).put(76);
        allocate.flip();
        IntBuffer onlyBuffer = allocate.asReadOnlyBuffer();
        System.out.println(onlyBuffer.hasArray());
        System.out.println(onlyBuffer.get());
        System.out.println(onlyBuffer.isReadOnly());
        System.out.println(allocate);
        System.out.println(onlyBuffer);
    }
}
```

执行结果如下：

```
false
56
true
java.nio.HeapIntBuffer[pos = 0 lim = 3 cap = 1024]
java.nio.HeapIntBufferR[pos = 1 lim = 3 cap = 1024]
```

7. clear()

重置当前缓冲区内的数据指针。

8. compact()

压缩当前缓冲区的数据并切换为输入模式。

9. compareTo(IntBuffer that)

比较当前缓冲区和指定的缓冲区数据是否相同,返回 int 值。接收 IntBuffer 入参,作为指定的缓冲区。

10. duplicate()

返回当前缓冲区对象的浅复制。

11. equals(Object ob)

比较当前对象和指定的对象,返回 boolean 值。接收 Object 入参,作为指定的对象,代码如下:

```
//第 10 章/four/IntBufferTest.java
public class IntBufferTest {

    public static void main(String[] args) {
        IntBuffer allocate = IntBuffer.allocate(1024);
        allocate.put(56).put(66).put(76);
        IntBuffer allocateA = IntBuffer.allocate(1024);
        allocateA.put(36).put(66).put(76);
        allocate.flip();
        allocateA.flip();
        System.out.println(allocate.equals(allocateA));
    }
}
```

执行结果如下:

```
false
```

12. flip()

将当前缓冲区的模式切换为输出模式,代码如下:

```
//第 10 章/four/IntBufferTest.java
public class IntBufferTest {

    public static void main(String[] args) {
        IntBuffer allocate = IntBuffer.allocate(1024);
        System.out.println(allocate);
        allocate.put(56).put(66).put(76);
        System.out.println(allocate);
        allocate.flip();
        System.out.println(allocate.get());
        System.out.println(allocate);
    }
}
```

执行结果如下:

```
java.nio.HeapIntBuffer[pos = 0 lim = 1024 cap = 1024]
java.nio.HeapIntBuffer[pos = 3 lim = 1024 cap = 1024]
56
java.nio.HeapIntBuffer[pos = 1 lim = 3 cap = 1024]
```

13. get()

返回当前缓冲区的整数数据。如果没有数据，则抛出异常。

14. get(int index)

返回当前缓冲区指定索引的整数数据。如果索引超标，则抛出异常。接收 int 入参，作为指定索引，代码如下：

```
//第 10 章/four/IntBufferTest.java
public class IntBufferTest {

    public static void main(String[] args) {
        IntBuffer allocate = IntBuffer.allocate(1024);
        System.out.println(allocate);
        allocate.put(56).put(66).put(76);
        System.out.println(allocate);
        allocate.flip();
        System.out.println(allocate.get(1));
        System.out.println(allocate);
    }
}
```

执行结果如下：

```
java.nio.HeapIntBuffer[pos = 0 lim = 1024 cap = 1024]
java.nio.HeapIntBuffer[pos = 3 lim = 1024 cap = 1024]
66
java.nio.HeapIntBuffer[pos = 0 lim = 3 cap = 1024]
```

15. get(int[] dst)

将当前缓冲区的整数数据存储到指定的整数数组中。接收 int[]入参，作为指定的整数数组，代码如下：

```
//第 10 章/four/IntBufferTest.java
public class IntBufferTest {

    public static void main(String[] args) {
        IntBuffer allocate = IntBuffer.allocate(1024);
        System.out.println(allocate);
        allocate.put(56).put(66).put(76);
        System.out.println(allocate);
        allocate.flip();
        int[] ints = new int[allocate.remaining()];
        System.out.println(allocate.get(ints));
        System.out.println(allocate);
        System.out.println(Arrays.toString(ints));
    }
}
```

执行结果如下：

```
java.nio.HeapIntBuffer[pos = 0 lim = 1024 cap = 1024]
java.nio.HeapIntBuffer[pos = 3 lim = 1024 cap = 1024]
java.nio.HeapIntBuffer[pos = 3 lim = 3 cap = 1024]
java.nio.HeapIntBuffer[pos = 3 lim = 3 cap = 1024]
[56, 66, 76]
```

16. get(int[] dst, int offset, int length)

将当前缓冲区的整数数据存储到指定的整数数组中。接收 int[]入参,作为指定的整数数组,接收 int 入参,作为数组的起始索引,接收 int 入参,作为最大的数据长度。

17. get(int index, int[] dst)

将当前缓冲区的整数数据存储到指定的整数数组中。接收 int 入参,作为缓冲区的起始索引,接收 int[]入参,作为指定的整数数组。

18. hasArray()

检查当前缓冲区是否由可访问的整数数组支持,返回 boolean 值。

19. isDirect()

检查当前缓冲区是否为直接缓冲区,返回 boolean 值。

20. isEmpty()

检查当前缓冲区数据是否为空,返回 boolean 值。

21. limit(int newLimit)

设置当前缓冲区数据指针的位置限制。接收 int 入参,作为数据指针的位置限制。

22. mark()

设置当前缓冲区的数据指针标记。

23. mismatch(IntBuffer that)

返回当前缓冲区与指定缓冲区之间第 1 个不匹配数据的相对索引。接收 IntBuffer 入参,作为指定缓冲区。

24. order()

返回当前缓冲区的排序规则。

25. position(int newPosition)

设置当前缓冲区数据指针的位置。接收 int 入参,作为数据指针的位置。

26. position()

返回当前缓冲区数据指针的位置。

27. put(int i)

添加当前缓冲区的整数数据。接收 int 入参,作为整数数据。

28. put(int[] src)

添加当前缓冲区的整数数组数据。接收 int[]入参，作为整数数组数据。

29. put(int[] src, int offset, int length)

添加当前缓冲区的整数数组数据。接收 int[]入参，作为整数数组数据，接收 int 入参，作为数组的起始索引，接收 int 入参，作为最大的数据长度。

30. put(int index, int i)

修改当前缓冲区指定索引的整数数据。接收 int 入参，作为指定索引，接收 int 入参，作为整数数据，代码如下：

```
//第 10 章/four/IntBufferTest.java
public class IntBufferTest {

    public static void main(String[] args) {
        IntBuffer allocate = IntBuffer.allocate(1024);
        System.out.println(allocate);
        allocate.put(56).put(66).put(76);
        allocate.put(1,666);
        System.out.println(allocate);
        allocate.flip();
        int[] ints = new int[allocate.remaining()];
        System.out.println(allocate.get(ints));
        System.out.println(allocate);
        System.out.println(Arrays.toString(ints));
    }
}
```

执行结果如下：

```
java.nio.HeapIntBuffer[pos = 0 lim = 1024 cap = 1024]
java.nio.HeapIntBuffer[pos = 3 lim = 1024 cap = 1024]
java.nio.HeapIntBuffer[pos = 3 lim = 3 cap = 1024]
java.nio.HeapIntBuffer[pos = 3 lim = 3 cap = 1024]
[56, 666, 76]
```

31. put(int index, int[] src)

修改当前缓冲区指定索引的整数数组数据。接收 int 入参，作为指定索引，接收 int[]入参，作为整数数组数据。

32. reset()

将当前缓冲区的数据指针重置为先前标记的值。

33. rewind()

将当前缓冲区的数据指针置为 0 并废弃标记值。

34. slice()

返回当前缓冲区对象的子序列。

35. slice(int index, int length)

返回当前缓冲区对象的子序列。接收 int 入参,作为数据起始索引,接收 int 入参,作为数据长度。

36. hasRemaining()

检查当前缓冲区位置与限制之间是否存在可用空间,返回 boolean 值。

37. remaining()

返回当前缓冲区位置与限制之间的元素数。

38. isReadOnly()

检查当前缓冲区是否为只读缓冲区,返回 boolean 值。

39. capacity()

返回当前缓冲区内底层封装数组的容量。

10.5 LongBuffer 抽象类

长整数缓冲区,核心字段见表 10-6。

表 10-6 核心字段

字 段	描 述
final long[] hb	长整数数组缓冲区
final int offset	数组索引的偏移量
boolean isReadOnly	是否为只读缓冲区

本节基于官方文档介绍常用方法,核心方法配合代码示例以方便读者理解。

1. allocate(int capacity)

静态方法,创建指定容量的长整数缓冲区对象。接收 int 入参,作为缓冲区的容量,代码如下:

```
//第 10 章/five/LongBufferTest.java
public class LongBufferTest {

    public static void main(String[] args) {
        LongBuffer allocate = LongBuffer.allocate(1024);
        System.out.println(allocate);
        allocate.put(56).put(66).put(76);
        System.out.println(allocate);
    }
}
```

执行结果如下：

```
java.nio.HeapLongBuffer[pos = 0 lim = 1024 cap = 1024]
java.nio.HeapLongBuffer[pos = 3 lim = 1024 cap = 1024]
```

2. wrap(long[] array)

静态方法，创建指定初始数据的长整数缓冲区对象。接收 long[]入参，作为缓冲区的初始数据，代码如下：

```
//第 10 章/five/LongBufferTest.java
public class LongBufferTest {

    public static void main(String[] args) {
        LongBuffer allocate = LongBuffer.wrap(new long[]{55,56,57});
        System.out.println(allocate);
        System.out.println(allocate.get());
        System.out.println(allocate.get());
        System.out.println(allocate.get());
        System.out.println(allocate);
    }
}
```

执行结果如下：

```
java.nio.HeapLongBuffer[pos = 0 lim = 3 cap = 3]
55
56
57
java.nio.HeapLongBuffer[pos = 3 lim = 3 cap = 3]
```

3. wrap(long[] array, int offset, int length)

静态方法，创建指定初始数据的长整数缓冲区对象。接收 long[]入参，作为缓冲区的初始数据，接收 int 入参，作为数组的起始索引，接收 int 入参，作为最大的数据长度。

4. array()

返回当前缓冲区内底层封装的长整数数组。

5. arrayOffset()

返回当前缓冲区内底层封装长整数数组的偏移量。

6. asReadOnlyBuffer()

创建新的只读长整数缓冲区并共享当前长整数缓冲区的内容，代码如下：

```
//第 10 章/five/LongBufferTest.java
public class LongBufferTest {

    public static void main(String[] args) {
        LongBuffer allocate = LongBuffer.allocate(1024);
```

```
        System.out.println(allocate);
        allocate.put(new long[]{55,56,57,58,59});
        System.out.println(allocate);
        allocate.flip();
        System.out.println(allocate);
        LongBuffer readOnlyBuffer = allocate.asReadOnlyBuffer();
        while (readOnlyBuffer.hasRemaining()){
            System.out.println(readOnlyBuffer.get());
        }
        System.out.println(allocate);
        System.out.println(readOnlyBuffer);
    }
}
```

执行结果如下：

```
java.nio.HeapLongBuffer[pos = 0 lim = 1024 cap = 1024]
java.nio.HeapLongBuffer[pos = 5 lim = 1024 cap = 1024]
java.nio.HeapLongBuffer[pos = 0 lim = 5 cap = 1024]
55
56
57
58
59
java.nio.HeapLongBuffer[pos = 0 lim = 5 cap = 1024]
java.nio.HeapLongBufferR[pos = 5 lim = 5 cap = 1024]
```

7. clear()

重置当前缓冲区内的数据指针。

8. compact()

压缩当前缓冲区的数据并切换为输入模式。

9. compareTo(LongBuffer that)

比较当前缓冲区和指定的缓冲区数据是否相同，返回 int 值。接收 LongBuffer 入参，作为指定的缓冲区。

10. duplicate()

返回当前缓冲区对象的浅复制。

11. equals(Object ob)

比较当前对象和指定的对象，返回 boolean 值。接收 Object 入参，作为指定的对象。

12. flip()

将当前缓冲区的模式切换为输出模式。

13. get()

返回当前缓冲区的长整数数据。如果没有数据，则抛出异常，代码如下：

```
//第 10 章/five/LongBufferTest.java
public class LongBufferTest {

    public static void main(String[] args) {
        LongBuffer allocate = LongBuffer.allocate(1024);
        System.out.println(allocate);
        allocate.put(new long[]{55,56,57,58,59});
        System.out.println(allocate);
        //将当前缓冲区的模式切换为输出模式
        allocate.flip();
        System.out.println(allocate);
        System.out.println(allocate.get());
        System.out.println(allocate.get());
        System.out.println(allocate.get());
        System.out.println(allocate);
        //压缩当前缓冲区的数据并切换为输入模式
        allocate.compact();
        System.out.println(allocate);
    }
}
```

执行结果如下：

```
java.nio.HeapLongBuffer[pos = 0 lim = 1024 cap = 1024]
java.nio.HeapLongBuffer[pos = 5 lim = 1024 cap = 1024]
java.nio.HeapLongBuffer[pos = 0 lim = 5 cap = 1024]
55
56
57
java.nio.HeapLongBuffer[pos = 3 lim = 5 cap = 1024]
java.nio.HeapLongBuffer[pos = 2 lim = 1024 cap = 1024]
```

14. get(int index)

返回当前缓冲区指定索引的长整数数据。如果索引超标，则抛出异常。接收 int 入参，作为指定索引，代码如下：

```
//第 10 章/five/LongBufferTest.java
public class LongBufferTest {

    public static void main(String[] args) {
        LongBuffer allocate = LongBuffer.allocate(1024);
        System.out.println(allocate);
        allocate.put(new long[]{55,56,57,58,59});
        System.out.println(allocate);
        //将当前缓冲区的模式切换为输出模式
        allocate.flip();
        System.out.println(allocate);
        System.out.println(allocate.get(0));
        System.out.println(allocate.get(1));
        System.out.println(allocate.get(2));
```

```
        System.out.println(allocate);
        //压缩当前缓冲区的数据并切换为输入模式
        allocate.compact();
        System.out.println(allocate);
    }
}
```

执行结果如下：

```
java.nio.HeapLongBuffer[pos = 0 lim = 1024 cap = 1024]
java.nio.HeapLongBuffer[pos = 5 lim = 1024 cap = 1024]
java.nio.HeapLongBuffer[pos = 0 lim = 5 cap = 1024]
55
56
57
java.nio.HeapLongBuffer[pos = 0 lim = 5 cap = 1024]
java.nio.HeapLongBuffer[pos = 5 lim = 1024 cap = 1024]
```

15. get(int index, long[] dst)

将当前缓冲区的长整数数据存储到指定的长整数数组中。接收 int 入参,作为缓冲区的起始索引,接收 long[]入参,作为指定的长整数数组。

16. get(long[] dst)

将当前缓冲区的长整数数据存储到指定的长整数数组中。接收 long[]入参,作为指定的长整数数组,代码如下：

```
//第 10 章/five/LongBufferTest.java
public class LongBufferTest {

    public static void main(String[] args) {
        LongBuffer allocate = LongBuffer.allocate(1024);
        System.out.println(allocate);
        allocate.put(new long[]{55,56,57,58,59});
        System.out.println(allocate);
        //将当前缓冲区的模式切换为输出模式
        allocate.flip();
        System.out.println(allocate);
        long[] longs = new long[allocate.remaining()];
        allocate.get(longs);
        System.out.println(Arrays.toString(longs));
        System.out.println(allocate);
        //压缩当前缓冲区的数据并切换为输入模式
        allocate.compact();
        System.out.println(allocate);
    }
}
```

执行结果如下：

```
java.nio.HeapLongBuffer[pos = 0 lim = 1024 cap = 1024]
java.nio.HeapLongBuffer[pos = 5 lim = 1024 cap = 1024]
java.nio.HeapLongBuffer[pos = 0 lim = 5 cap = 1024]
[55, 56, 57, 58, 59]
java.nio.HeapLongBuffer[pos = 5 lim = 5 cap = 1024]
java.nio.HeapLongBuffer[pos = 0 lim = 1024 cap = 1024]
```

17. get(long[] dst, int offset, int length)

将当前缓冲区的长整数数据存储到指定的长整数数组中。接收 long[]入参，作为指定的长整数数组，接收 int 入参，作为数组的起始索引，接收 int 入参，作为最大的数据长度。

18. hasArray()

检查当前缓冲区是否由可访问的长整数数组支持，返回 boolean 值。

19. isDirect()

检查当前缓冲区是否为直接缓冲区，返回 boolean 值。

20. isEmpty()

检查当前缓冲区数据是否为空，返回 boolean 值。

21. limit(int newLimit)

设置当前缓冲区数据指针的位置限制。接收 int 入参，作为数据指针的位置限制。

22. mark()

设置当前缓冲区的数据指针标记。

23. mismatch(LongBuffer that)

返回当前缓冲区与指定缓冲区之间第 1 个不匹配数据的相对索引。接收 LongBuffer 入参，作为指定缓冲区。

24. order()

返回当前缓冲区的排序规则。

25. position(int newPosition)

设置当前缓冲区数据指针的位置。接收 int 入参，作为数据指针的位置。

26. put(int index, long l)

修改当前缓冲区指定索引的长整数数据。接收 int 入参，作为指定索引，接收 long 入参，作为长整数数据，代码如下：

```
//第 10 章/five/LongBufferTest.java
public class LongBufferTest {

    public static void main(String[] args) {
        LongBuffer allocate = LongBuffer.allocate(1024);
```

```
        System.out.println(allocate);
        allocate.put(new long[]{55,56,57,58,59});
        System.out.println(allocate);
        allocate.put(2,66666);
        //将当前缓冲区的模式切换为输出模式
        allocate.flip();
        System.out.println(allocate);
        long[] longs = new long[allocate.remaining()];
        allocate.get(longs);
        System.out.println(Arrays.toString(longs));
        System.out.println(allocate);
        //压缩当前缓冲区的数据并切换为输入模式
        allocate.compact();
        System.out.println(allocate);
    }
}
```

执行结果如下：

```
java.nio.HeapLongBuffer[pos=0 lim=1024 cap=1024]
java.nio.HeapLongBuffer[pos=5 lim=1024 cap=1024]
java.nio.HeapLongBuffer[pos=0 lim=5 cap=1024]
[55, 56, 66666, 58, 59]
java.nio.HeapLongBuffer[pos=5 lim=5 cap=1024]
java.nio.HeapLongBuffer[pos=0 lim=1024 cap=1024]
```

27. put(int index, long[] src)

修改当前缓冲区指定索引的长整数数组数据。接收 int 入参，作为指定索引，接收 long[]入参，作为长整数数组数据。

28. put(long l)

添加当前缓冲区的长整数数据。接收 long 入参，作为长整数数据。

29. put(long[] src)

添加当前缓冲区的长整数数组数据。接收 long[]入参，作为长整数数组数据。

30. put(long[] src, int offset, int length)

添加当前缓冲区的长整数数组数据。接收 long[]入参，作为长整数数组数据，接收 int 入参，作为数组的起始索引，接收 int 入参，作为最大的数据长度。

31. reset()

将当前缓冲区的数据指针重置为先前标记的值。

32. rewind()

将当前缓冲区的数据指针置为 0 并废弃标记值。

33. slice()

返回当前缓冲区对象的子序列。

34. slice(int index, int length)

返回当前缓冲区对象的子序列。接收 int 入参，作为数据起始索引，接收 int 入参，作为数据长度。

35. hasRemaining()

检查当前缓冲区位置与限制之间是否存在可用空间，返回 boolean 值。

36. remaining()

返回当前缓冲区位置与限制之间的元素数。

37. isReadOnly()

检查当前缓冲区是否为只读缓冲区，返回 boolean 值。

38. capacity()

返回当前缓冲区内底层封装数组的容量。

10.6 ShortBuffer 抽象类

短整数缓冲区，核心字段见表 10-7。

表 10-7 核心字段

字　段	描　述
final short[] hb	短整数数组缓冲区
final int offset	数组索引的偏移量
boolean isReadOnly	是否为只读缓冲区

本节基于官方文档介绍常用方法，核心方法配合代码示例以方便读者理解。

1. allocate(int capacity)

静态方法，创建指定容量的短整数缓冲区对象。接收 int 入参，作为缓冲区的容量，代码如下：

```
//第 10 章/six/ShortBufferTest.java
public class ShortBufferTest {

   public static void main(String[] args) {
      ShortBuffer allocate = ShortBuffer.allocate(1024);
      System.out.println(allocate);
      allocate.put((short) 56).put((short) 66).put((short) 76);
      System.out.println(allocate);
      allocate.flip();
      System.out.println(allocate);
   }
}
```

执行结果如下：

```
java.nio.HeapShortBuffer[pos = 0 lim = 1024 cap = 1024]
java.nio.HeapShortBuffer[pos = 3 lim = 1024 cap = 1024]
java.nio.HeapShortBuffer[pos = 0 lim = 3 cap = 1024]
```

2. wrap(short[] array)

静态方法，创建指定初始数据的短整数缓冲区对象。接收 short[]入参，作为缓冲区的初始数据，代码如下：

```java
//第 10 章/six/ShortBufferTest.java
public class ShortBufferTest {

    public static void main(String[] args) {
        ShortBuffer allocate = ShortBuffer.wrap(new short[]{55,56,57});
        System.out.println(allocate);
        while (allocate.hasRemaining()){
            System.out.println(allocate.get());
        }
        System.out.println(allocate);
    }
}
```

执行结果如下：

```
java.nio.HeapShortBuffer[pos = 0 lim = 3 cap = 3]
55
56
57
java.nio.HeapShortBuffer[pos = 3 lim = 3 cap = 3]
```

3. wrap(short[] array, int offset, int length)

静态方法，创建指定初始数据的短整数缓冲区对象。接收 short[]入参，作为缓冲区的初始数据，接收 int 入参，作为数组的起始索引，接收 int 入参，作为最大的数据长度。

4. array()

返回当前缓冲区内底层封装的短整数数组。

5. arrayOffset()

返回当前缓冲区内底层封装短整数数组的偏移量。

6. asReadOnlyBuffer()

创建新的只读短整数缓冲区并共享当前短整数缓冲区的内容。

7. clear()

重置当前缓冲区内的数据指针。

8．compact()

压缩当前缓冲区的数据并切换为输入模式。

9．compareTo(ShortBuffer that)

比较当前缓冲区和指定的缓冲区数据是否相同，返回 int 值。接收 ShortBuffer 入参，作为指定的缓冲区。

10．duplicate()

返回当前缓冲区对象的浅复制。

11．equals(Object ob)

比较当前对象和指定的对象，返回 boolean 值。接收 Object 入参，作为指定的对象。

12．flip()

将当前缓冲区的模式切换为输出模式。

13．get()

返回当前缓冲区的短整数数据。如果没有数据，则抛出异常，代码如下：

```
//第 10 章/six/ShortBufferTest.java
public class ShortBufferTest {

    public static void main(String[] args) {
        ShortBuffer allocate = ShortBuffer.allocate(1024);
        System.out.println(allocate);
        System.out.println(allocate.arrayOffset());
        allocate.put((short) 56).put((short) 66).put((short) 76);
        allocate.flip();
        while (allocate.hasRemaining()){
            System.out.println(allocate.get());
        }
        System.out.println(allocate);
        allocate.compact();
        System.out.println(allocate);
    }
}
```

执行结果如下：

```
java.nio.HeapShortBuffer[pos = 0 lim = 1024 cap = 1024]
0
56
66
76
java.nio.HeapShortBuffer[pos = 3 lim = 3 cap = 1024]
java.nio.HeapShortBuffer[pos = 0 lim = 1024 cap = 1024]
```

14. get(int index)

返回当前缓冲区指定索引的短整数数据。如果索引超标，则抛出异常。接收 int 入参，作为指定索引，代码如下：

```
//第 10 章/six/ShortBufferTest.java
public class ShortBufferTest {

    public static void main(String[] args) {
        ShortBuffer allocate = ShortBuffer.allocate(1024);
        System.out.println(allocate);
        System.out.println(allocate.arrayOffset());
        allocate.put((short) 56).put((short) 66).put((short) 76);
        allocate.flip();
        System.out.println(allocate.get(0));
        System.out.println(allocate.get(1));
        System.out.println(allocate.get(2));
        System.out.println(allocate);
        allocate.compact();
        System.out.println(allocate);
    }
}
```

执行结果如下：

```
java.nio.HeapShortBuffer[pos = 0 lim = 1024 cap = 1024]
0
56
66
76
java.nio.HeapShortBuffer[pos = 0 lim = 3 cap = 1024]
java.nio.HeapShortBuffer[pos = 3 lim = 1024 cap = 1024]
```

15. get(int index, short[] dst)

将当前缓冲区的短整数数据存储到指定的短整数数组中。接收 int 入参，作为缓冲区的起始索引，接收 short[]入参，作为指定的短整数数组。

16. get(short[] dst)

将当前缓冲区的短整数数据存储到指定的短整数数组中。接收 short[]入参，作为指定的短整数数组。

17. get(short[] dst, int offset, int length)

将当前缓冲区的短整数数据存储到指定的短整数数组中。接收 short[]入参，作为指定的短整数数组，接收 int 入参，作为数组的起始索引，接收 int 入参，作为最大的数据长度。

18. hasArray()

检查当前缓冲区是否由可访问的短整数数组支持，返回 boolean 值。

19. isDirect()

检查当前缓冲区是否为直接缓冲区,返回 boolean 值。

20. isEmpty()

检查当前缓冲区数据是否为空,返回 boolean 值。

21. limit(int newLimit)

设置当前缓冲区数据指针的位置限制。接收 int 入参,作为数据指针的位置限制。

22. mark()

设置当前缓冲区的数据指针标记。

23. mismatch(ShortBuffer that)

返回当前缓冲区与指定缓冲区之间第 1 个不匹配数据的相对索引。接收 ShortBuffer 入参,作为指定缓冲区。

24. order()

返回当前缓冲区的排序规则。

25. position(int newPosition)

设置当前缓冲区数据指针的位置。接收 int 入参,作为数据指针的位置。

26. put(int index, short s)

修改当前缓冲区指定索引的短整数数据。接收 int 入参,作为指定索引,接收 short 入参,作为短整数数据,代码如下:

```
//第 10 章/six/ShortBufferTest.java
public class ShortBufferTest {

    public static void main(String[] args) {
        ShortBuffer allocate = ShortBuffer.allocate(1024);
        System.out.println(allocate);
        System.out.println(allocate.arrayOffset());
        allocate.put((short) 56).put((short) 66).put((short) 76);
        allocate.put(1, (short) 6666);
        allocate.flip();
        System.out.println(allocate.get(0));
        System.out.println(allocate.get(1));
        System.out.println(allocate.get(2));
        System.out.println(allocate);
        allocate.compact();
        System.out.println(allocate);
    }
}
```

执行结果如下:

```
java.nio.HeapShortBuffer[pos = 0 lim = 1024 cap = 1024]
0
56
6666
76
java.nio.HeapShortBuffer[pos = 0 lim = 3 cap = 1024]
java.nio.HeapShortBuffer[pos = 3 lim = 1024 cap = 1024]
```

27. put(int index, short[] src)

修改当前缓冲区指定索引的短整数数组数据。接收 int 入参,作为指定索引,接收 short[]入参,作为短整数数组数据。

28. put(short s)

添加当前缓冲区的短整数数据。接收 short 入参,作为短整数数据。

29. put(short[] src)

添加当前缓冲区的短整数数组数据。接收 short[]入参,作为短整数数组数据。

30. put(short[] src, int offset, int length)

添加当前缓冲区的短整数数组数据。接收 short[]入参,作为短整数数组数据,接收 int 入参,作为数组的起始索引,接收 int 入参,作为最大的数据长度。

31. reset()

将当前缓冲区的数据指针重置为先前标记的值。

32. rewind()

将当前缓冲区的数据指针置为 0 并废弃标记值。

33. slice()

返回当前缓冲区对象的子序列。

34. slice(int index, int length)

返回当前缓冲区对象的子序列。接收 int 入参,作为数据起始索引,接收 int 入参,作为数据长度。

35. hasRemaining()

检查当前缓冲区位置与限制之间是否存在可用空间,返回 boolean 值。

36. remaining()

返回当前缓冲区位置与限制之间的元素数。

37. isReadOnly()

检查当前缓冲区是否为只读缓冲区,返回 boolean 值。

38. capacity()

返回当前缓冲区内底层封装数组的容量。

10.7 FloatBuffer 抽象类

单精度浮点数缓冲区，核心字段见表10-8。

表10-8 核心字段

字 段	描 述
final float[] hb	单精度浮点数数组缓冲区
final int offset	数组索引的偏移量
boolean isReadOnly	是否为只读缓冲区

本节基于官方文档介绍常用方法，核心方法配合代码示例以方便读者理解。

1. allocate(int capacity)

静态方法，创建指定容量的单精度浮点数缓冲区对象。接收int入参，作为缓冲区的容量，代码如下：

```
//第10章/seven/FloatBufferTest.java
public class FloatBufferTest {

    public static void main(String[] args) {
        FloatBuffer allocate = FloatBuffer.allocate(1024);
        System.out.println(allocate);
        allocate.put(new float[]{55,56,57,58});
        System.out.println(allocate);
        allocate.flip();
        System.out.println(allocate);
    }
}
```

执行结果如下：

```
java.nio.HeapFloatBuffer[pos = 0 lim = 1024 cap = 1024]
java.nio.HeapFloatBuffer[pos = 4 lim = 1024 cap = 1024]
java.nio.HeapFloatBuffer[pos = 0 lim = 4 cap = 1024]
```

2. wrap(float[] array)

静态方法，创建指定初始数据的单精度浮点数缓冲区对象。接收float[]入参，作为缓冲区的初始数据，代码如下：

```
//第10章/seven/FloatBufferTest.java
public class FloatBufferTest {

    public static void main(String[] args) {
        FloatBuffer allocate = FloatBuffer.wrap(
                                        new float[]{55,56,57,58,59});
        System.out.println(allocate);
```

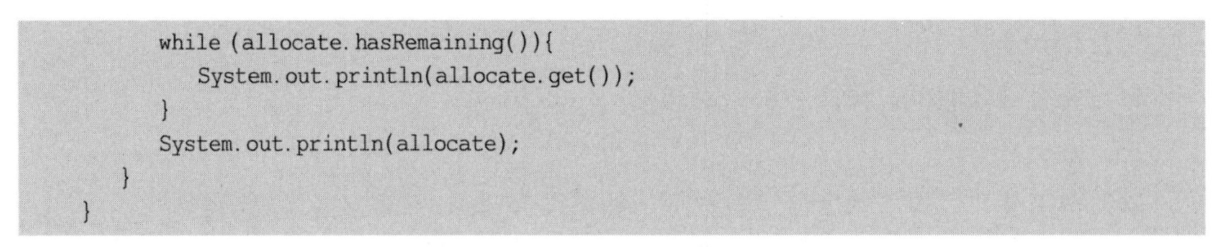

```
        while (allocate.hasRemaining()){
            System.out.println(allocate.get());
        }
        System.out.println(allocate);
    }
}
```

执行结果如下：

```
java.nio.HeapFloatBuffer[pos = 0 lim = 5 cap = 5]
55.0
56.0
57.0
58.0
59.0
java.nio.HeapFloatBuffer[pos = 5 lim = 5 cap = 5]
```

3. wrap(float[] array, int offset, int length)

静态方法,创建指定初始数据的单精度浮点数缓冲区对象。接收 float[]入参,作为缓冲区的初始数据,接收 int 入参,作为数组的起始索引,接收 int 入参,作为最大的数据长度。

4. array()

返回当前缓冲区内底层封装的单精度浮点数数组。

5. arrayOffset()

返回当前缓冲区内底层封装单精度浮点数数组的偏移量。

6. asReadOnlyBuffer()

创建新的只读单精度浮点数缓冲区并共享当前单精度浮点数缓冲区的内容。

7. clear()

重置当前缓冲区内的数据指针。

8. compact()

压缩当前缓冲区的数据并切换为输入模式。

9. compareTo(FloatBuffer that)

比较当前缓冲区和指定的缓冲区数据是否相同,返回 int 值。接收 FloatBuffer 入参,作为指定的缓冲区。

10. duplicate()

返回当前缓冲区对象的浅复制。

11. equals(Object ob)

比较当前对象和指定的对象,返回 boolean 值。接收 Object 入参,作为指定的对象。

12. flip()

将当前缓冲区的模式切换为输出模式。

13. get()

返回当前缓冲区的单精度浮点数数据。如果没有数据，则抛出异常，代码如下：

```
//第 10 章/seven/FloatBufferTest.java
public class FloatBufferTest {

    public static void main(String[] args) {
        FloatBuffer allocate = FloatBuffer.allocate(1024);
        System.out.println(allocate);
        allocate.put(new float[]{55,56,57,58});
        System.out.println(allocate);
        allocate.flip();
        while (allocate.hasRemaining()){
            System.out.println(allocate.get());
        }
        System.out.println(allocate);
    }
}
```

执行结果如下：

```
java.nio.HeapFloatBuffer[pos = 0 lim = 1024 cap = 1024]
java.nio.HeapFloatBuffer[pos = 4 lim = 1024 cap = 1024]
55.0
56.0
57.0
58.0
java.nio.HeapFloatBuffer[pos = 4 lim = 4 cap = 1024]
```

14. get(float[] dst)

将当前缓冲区的单精度浮点数数据存储到指定的单精度浮点数数组中。接收 float[] 入参，作为指定的单精度浮点数数组，代码如下：

```
//第 10 章/seven/FloatBufferTest.java
public class FloatBufferTest {

    public static void main(String[] args) {
        FloatBuffer allocate = FloatBuffer.allocate(1024);
        System.out.println(allocate);
        allocate.put(new float[]{55,56,57,58});
        System.out.println(allocate);
        allocate.flip();
        float[] fs = new float[allocate.remaining()];
        allocate.get(fs);
        System.out.println(Arrays.toString(fs));
        System.out.println(allocate);
    }
}
```

执行结果如下：

```
java.nio.HeapFloatBuffer[pos = 0 lim = 1024 cap = 1024]
java.nio.HeapFloatBuffer[pos = 4 lim = 1024 cap = 1024]
[55.0, 56.0, 57.0, 58.0]
java.nio.HeapFloatBuffer[pos = 4 lim = 4 cap = 1024]
```

15. get(float[] dst，int offset，int length)

将当前缓冲区的单精度浮点数数据存储到指定的单精度浮点数数组中。接收 float[]入参，作为指定的单精度浮点数数组，接收 int 入参，作为数组的起始索引，接收 int 入参，作为最大的数据长度。

16. get(int index)

返回当前缓冲区指定索引的单精度浮点数数据。如果索引超标，则抛出异常。接收 int 入参，作为指定索引，代码如下：

```
//第 10 章/seven/FloatBufferTest.java
public class FloatBufferTest {

    public static void main(String[] args) {
        FloatBuffer allocate = FloatBuffer.allocate(1024);
        System.out.println(allocate);
        allocate.put(new float[]{55,56,57,58});
        System.out.println(allocate);
        allocate.flip();
        for (int i = 0; i < allocate.remaining(); i++) {
            System.out.println(allocate.get(i));
        }
        System.out.println(allocate);
    }
}
```

执行结果如下：

```
java.nio.HeapFloatBuffer[pos = 0 lim = 1024 cap = 1024]
java.nio.HeapFloatBuffer[pos = 4 lim = 1024 cap = 1024]
55.0
56.0
57.0
58.0
java.nio.HeapFloatBuffer[pos = 0 lim = 4 cap = 1024]
```

17. get(int index，float[] dst)

将当前缓冲区的单精度浮点数数据存储到指定的单精度浮点数数组中。接收 int 入参，作为缓冲区的起始索引，接收 float[]入参，作为指定的单精度浮点数数组。

18. hasArray()

检查当前缓冲区是否由可访问的单精度浮点数数组支持，返回 boolean 值。

19. isDirect()

检查当前缓冲区是否为直接缓冲区，返回 boolean 值。

20. isEmpty()

检查当前缓冲区数据是否为空，返回 boolean 值。

21. limit(int newLimit)

设置当前缓冲区数据指针的位置限制。接收 int 入参，作为数据指针的位置限制。

22. mark()

设置当前缓冲区的数据指针标记。

23. mismatch(FloatBuffer that)

返回当前缓冲区与指定缓冲区之间第 1 个不匹配数据的相对索引。接收 FloatBuffer 入参，作为指定缓冲区。

24. order()

返回当前缓冲区的排序规则。

25. position(int newPosition)

设置当前缓冲区数据指针的位置。接收 int 入参，作为数据指针的位置。

26. put(float f)

添加当前缓冲区的单精度浮点数数据。接收 float 入参，作为单精度浮点数数据。

27. put(float[] src)

添加当前缓冲区的单精度浮点数组数据。接收 float[]入参，作为单精度浮点数数组数据，代码如下：

```
//第 10 章/seven/FloatBufferTest.java
public class FloatBufferTest {

    public static void main(String[] args) {
        FloatBuffer allocate = FloatBuffer.allocate(1024);
        System.out.println(allocate);
        allocate.put(new float[]{55,56,57,58});
        System.out.println(allocate);
        allocate.flip();
        int remaining = allocate.remaining();
        for (int i = 0; i < remaining; i++) {
            System.out.println(allocate.get());
        }
        System.out.println(allocate);
    }
}
```

执行结果如下：

```
java.nio.HeapFloatBuffer[pos = 0 lim = 1024 cap = 1024]
java.nio.HeapFloatBuffer[pos = 4 lim = 1024 cap = 1024]
55.0
56.0
57.0
58.0
java.nio.HeapFloatBuffer[pos = 4 lim = 4 cap = 1024]
```

28. put(float[] src, int offset, int length)

添加当前缓冲区的单精度浮点数数组数据。接收 float[]入参,作为单精度浮点数数组数据,接收 int 入参,作为数组的起始索引,接收 int 入参,作为最大的数据长度。

29. put(int index, float f)

修改当前缓冲区指定索引的单精度浮点数数据。接收 int 入参,作为指定索引,接收 float 入参,作为单精度浮点数数据。

30. put(int index, float[] src)

修改当前缓冲区指定索引的单精度浮点数数组数据。接收 int 入参,作为指定索引,接收 float[]入参,作为单精度浮点数数组数据。

31. put(FloatBuffer src)

添加当前缓冲区的单精度浮点数数据。接收 FloatBuffer 入参,作为单精度浮点数数据,代码如下:

```
//第 10 章/seven/FloatBufferTest.java
public class FloatBufferTest {

    public static void main(String[] args) {
        FloatBuffer allocate = FloatBuffer.allocate(1024);
        System.out.println(allocate);
        allocate.put(new float[]{55,56,57,58});
        System.out.println(allocate);
        allocate.flip();
        FloatBuffer ano = FloatBuffer.allocate(allocate.remaining());
        ano.put(allocate);
        ano.flip();
        float[] floats = new float[ano.remaining()];
        ano.get(floats);
        System.out.println(Arrays.toString(floats));
        System.out.println(ano);
        System.out.println(allocate);
    }
}
```

执行结果如下:

```
java.nio.HeapFloatBuffer[pos = 0 lim = 1024 cap = 1024]
java.nio.HeapFloatBuffer[pos = 4 lim = 1024 cap = 1024]
[55.0, 56.0, 57.0, 58.0]
java.nio.HeapFloatBuffer[pos = 4 lim = 4 cap = 4]
java.nio.HeapFloatBuffer[pos = 4 lim = 4 cap = 1024]
```

32. reset()

将当前缓冲区的数据指针重置为先前标记的值。

33. rewind()

将当前缓冲区的数据指针置为 0 并废弃标记值。

34. slice()

返回当前缓冲区对象的子序列。

35. slice(int index, int length)

返回当前缓冲区对象的子序列。接收 int 入参,作为数据起始索引,接收 int 入参,作为数据长度。

36. hasRemaining()

检查当前缓冲区位置与限制之间是否存在可用空间,返回 boolean 值。

37. remaining()

返回当前缓冲区位置与限制之间的元素数。

38. isReadOnly()

检查当前缓冲区是否为只读缓冲区,返回 boolean 值。

39. capacity()

返回当前缓冲区内底层封装数组的容量。

10.8 DoubleBuffer 抽象类

双精度浮点数缓冲区,核心字段见表 10-9。

表 10-9 核心字段

字 段	描 述
final double[] hb	双精度浮点数数组缓冲区
final int offset	数组索引的偏移量
boolean isReadOnly	是否为只读缓冲区

本节基于官方文档介绍常用方法,核心方法配合代码示例以方便读者理解。

1. allocate(int capacity)

静态方法,创建指定容量的双精度浮点数缓冲区对象。接收 int 入参,作为缓冲区的容量，代码如下：

```
//第 10 章/eight/DoubleBufferTest.java
public class DoubleBufferTest {

    public static void main(String[] args) {
        DoubleBuffer allocate = DoubleBuffer.allocate(1024);
        System.out.println(allocate);

        DoubleBuffer wrap = DoubleBuffer.wrap(new double[]
                                        {55, 56, 57, 58, 59});
        System.out.println(wrap);
    }
}
```

执行结果如下：

```
java.nio.HeapDoubleBuffer[pos = 0 lim = 1024 cap = 1024]
java.nio.HeapDoubleBuffer[pos = 0 lim = 5 cap = 5]
```

2. wrap(double[] array)

静态方法，创建指定初始数据的双精度浮点数缓冲区对象。接收 double[]入参，作为缓冲区的初始数据。

3. wrap(double[] array, int offset, int length)

静态方法，创建指定初始数据的双精度浮点数缓冲区对象。接收 double[]入参，作为缓冲区的初始数据，接收 int 入参，作为数组的起始索引，接收 int 入参，作为最大的数据长度。

4. array()

返回当前缓冲区内底层封装的双精度浮点数数组。

5. arrayOffset()

返回当前缓冲区内底层封装双精度浮点数数组的偏移量。

6. asReadOnlyBuffer()

创建新的只读双精度浮点数缓冲区并共享当前单精度浮点数缓冲区的内容。

7. clear()

重置当前缓冲区内的数据指针。

8. compact()

压缩当前缓冲区的数据并切换为输入模式。

9. compareTo(FloatBuffer that)

比较当前缓冲区和指定的缓冲区数据是否相同，返回 int 值。接收 FloatBuffer 入参，作为指定的缓冲区。

10. duplicate()

返回当前缓冲区对象的浅复制。

11. equals(Object ob)

比较当前对象和指定的对象，返回 boolean 值。接收 Object 入参，作为指定的对象。

12. flip()

将当前缓冲区的模式切换为输出模式。

13. get()

返回当前缓冲区的双精度浮点数数据。如果没有数据，则抛出异常。

14. get(double[] dst)

将当前缓冲区的双精度浮点数数据存储到指定的双精度浮点数数组中。接收 double[]入参，作为指定的双精度浮点数数组，代码如下：

```
//第 10 章/eight/DoubleBufferTest.java
public class DoubleBufferTest {

    public static void main(String[] args) {
        DoubleBuffer allocate = DoubleBuffer.allocate(1024);
        System.out.println(allocate);
        allocate.put(56).put(66).put(76).put(86).put(96);
        allocate.put(2,66666);
        allocate.flip();
        double[] doubles = new double[allocate.remaining()];
        allocate.get(doubles);
        System.out.println(Arrays.toString(doubles));
        System.out.println(allocate);
        allocate.compact();
        System.out.println(allocate);
    }
}
```

执行结果如下：

```
java.nio.HeapDoubleBuffer[pos=0 lim=1024 cap=1024]
[56.0, 66.0, 66666.0, 86.0, 96.0]
java.nio.HeapDoubleBuffer[pos=5 lim=5 cap=1024]
java.nio.HeapDoubleBuffer[pos=0 lim=1024 cap=1024]
```

15. get(double[] dst, int offset, int length)

将当前缓冲区的双精度浮点数数据存储到指定的双精度浮点数数组中。接收 double[]入参，作为指定的双精度浮点数数组，接收 int 入参，作为数组的起始索引，接收 int 入参，作为最大的数据长度。

16. get(int index)

返回当前缓冲区指定索引的双精度浮点数数据。如果索引超标，则抛出异常。接收 int 入参，作为指定索引。

17. get(int index, double[] dst)

将当前缓冲区的双精度浮点数数据存储到指定的双精度浮点数数组中。接收 int 入参，作为缓冲区的起始索引，接收 double[]入参，作为指定的双精度浮点数数组，代码如下：

```java
//第 10 章/eight/DoubleBufferTest.java
public class DoubleBufferTest {

    public static void main(String[] args) {
        DoubleBuffer allocate = DoubleBuffer.allocate(1024);
        System.out.println(allocate);
        allocate.put(56).put(66).put(76).put(86).put(96);
        allocate.put(2, 66666);
        allocate.flip();
        double[] doubles = new double[allocate.remaining()];
        allocate.get(0, doubles);
        System.out.println(Arrays.toString(doubles));
        System.out.println(allocate);
        allocate.compact();
        System.out.println(allocate);
    }
}
```

执行结果如下：

```
java.nio.HeapDoubleBuffer[pos = 0 lim = 1024 cap = 1024]
[56.0, 66.0, 66666.0, 86.0, 96.0]
java.nio.HeapDoubleBuffer[pos = 0 lim = 5 cap = 1024]
java.nio.HeapDoubleBuffer[pos = 5 lim = 1024 cap = 1024]
```

18. hasArray()

检查当前缓冲区是否由可访问的双精度浮点数数组支持，返回 boolean 值。

19. isDirect()

检查当前缓冲区是否为直接缓冲区，返回 boolean 值。

20. isEmpty()

检查当前缓冲区数据是否为空，返回 boolean 值。

21. limit(int newLimit)

设置当前缓冲区数据指针的位置限制。接收 int 入参，作为数据指针的位置限制。

22. mark()

设置当前缓冲区的数据指针标记。

23．mismatch(DoubleBuffer that)

返回当前缓冲区与指定缓冲区之间第 1 个不匹配数据的相对索引。接收 DoubleBuffer 入参，作为指定缓冲区。

24．order()

返回当前缓冲区的排序规则。

25．position(int newPosition)

设置当前缓冲区数据指针的位置。接收 int 入参，作为数据指针的位置。

26．put(double f)

添加当前缓冲区的双精度浮点数数据。接收 double 入参，作为双精度浮点数数据。

27．put(double[] src)

添加当前缓冲区的双精度浮点数数组数据。接收 double[]入参，作为双精度浮点数数组数据，代码如下：

```
//第 10 章/eight/DoubleBufferTest.java
public class DoubleBufferTest {

    public static void main(String[] args) {
        DoubleBuffer allocate = DoubleBuffer.allocate(1024);
        System.out.println(allocate);
        allocate.put(new double[]{55, 56, 57, 58, 59});
        allocate.flip();
        double[] doubles = new double[allocate.remaining()];
        allocate.get(doubles);
        System.out.println(Arrays.toString(doubles));
        System.out.println(allocate);
        allocate.compact();
        System.out.println(allocate);
    }
}
```

执行结果如下：

```
java.nio.HeapDoubleBuffer[pos = 0 lim = 1024 cap = 1024]
[55.0, 56.0, 57.0, 58.0, 59.0]
java.nio.HeapDoubleBuffer[pos = 5 lim = 5 cap = 1024]
java.nio.HeapDoubleBuffer[pos = 0 lim = 1024 cap = 1024]
```

28．put(double[] src, int offset, int length)

添加当前缓冲区的双精度浮点数数组数据。接收 double[]入参，作为双精度浮点数数组数据，接收 int 入参，作为数组的起始索引，接收 int 入参，作为最大的数据长度。

29．put(int index, double d)

修改当前缓冲区指定索引的双精度浮点数数据。接收 int 入参，作为指定索引，接收 double 入参，作为双精度浮点数数据。

30. put(int index, double[] src)

修改当前缓冲区指定索引的双精度浮点数数组数据。接收 int 入参,作为指定索引,接收 double[]入参,作为双精度浮点数数组数据。

31. put(DoubleBuffer src)

添加当前缓冲区的双精度浮点数数据。接收 DoubleBuffer 入参,作为双精度浮点数数据。

32. reset()

将当前缓冲区的数据指针重置为先前标记的值。

33. rewind()

将当前缓冲区的数据指针置为 0 并废弃标记值。

34. slice()

返回当前缓冲区对象的子序列。

35. slice(int index, int length)

返回当前缓冲区对象的子序列。接收 int 入参,作为数据起始索引,接收 int 入参,作为数据长度。

36. hasRemaining()

检查当前缓冲区位置与限制之间是否存在可用空间,返回 boolean 值。

37. remaining()

返回当前缓冲区位置与限制之间的元素数。

38. isReadOnly()

检查当前缓冲区是否为只读缓冲区,返回 boolean 值。

39. capacity()

返回当前缓冲区内底层封装数组的容量。

小结

本章介绍了特定基本类型的数据缓冲区,缓冲区是线性的有限集合。

习题

1. 判断题

(1) 缓冲区的实现类底层封装的都是数组类型。(　　)

(2) Buffer 抽象类的核心子类有 7 种类型。(　　)

(3) 每个缓冲区都有 capacity、limit、position、mark 这几个属性。(　　)

2. 选择题

(1) 返回缓冲区容量的方法是(　　)。(单选)

A. capacity()　　B. arrayOffset()

C. position()　　D. limit()

(2) 返回缓冲区当前数据指针位置的方法是(　　)。(单选)

A. capacity()　　B. arrayOffset()

C. position()　　D. limit()

(3) 返回缓冲区当前数据指针位置限制的方法是(　　)。(单选)

A. capacity()　　B. arrayOffset()

C. position()　　D. limit()

(4) 返回缓冲区底层封装数组的方法是(　　)。(单选)

A. array()　　B. arrayOffset()

C. position()　　D. limit()

(5) 表示字节缓冲区的类是(　　)。(单选)

A. ShortBuffer 类　　B. ByteBuffer 类

C. CharBuffer 类　　D. IntBuffer 类

3. 填空题

(1) 查看执行结果并补充代码,代码如下:

```
//第 10 章/answer/OneAnswer.java
public class OneAnswer {

    public static void main(String[] args) {
        //默认为输入模式
        ByteBuffer buffer = ByteBuffer.allocate(512);
        buffer.put("hello ByteBuffer ".getBytes());
        buffer.put("hello ByteBuffer".getBytes());
        System.out.println(buffer);
        buffer.________________;
        System.out.println(buffer);
        byte[] btsGet = new byte[buffer.________________];
        buffer.get(btsGet);
        System.out.println(new String(btsGet));
        System.out.println(buffer);
        buffer.________;
        System.out.println(buffer);
    }
}
```

执行结果如下:

```
java.nio.HeapByteBuffer[pos = 33 lim = 512 cap = 512]
java.nio.HeapByteBuffer[pos = 0 lim = 33 cap = 512]
hello ByteBuffer hello ByteBuffer
java.nio.HeapByteBuffer[pos = 33 lim = 33 cap = 512]
java.nio.HeapByteBuffer[pos = 0 lim = 512 cap = 512]
```

(2) 查看执行结果并补充代码,代码如下:

```
//第 10 章/answer/TwoAnswer.java
public class TwoAnswer {

    public static void main(String[] args) {
        ByteBuffer buffer = ByteBuffer.allocate(512);
        System.out.println(buffer);
        buffer.________________;
        System.out.println(buffer);
        buffer.________________;
        System.out.println(buffer);
        System.out.println(buffer.getLong());
        System.out.println(buffer);
        buffer.________________;
        System.out.println(buffer);
    }
}
```

执行结果如下:

```
java.nio.HeapByteBuffer[pos = 0 lim = 512 cap = 512]
java.nio.HeapByteBuffer[pos = 8 lim = 512 cap = 512]
java.nio.HeapByteBuffer[pos = 0 lim = 8 cap = 512]
15
java.nio.HeapByteBuffer[pos = 8 lim = 8 cap = 512]
java.nio.HeapByteBuffer[pos = 0 lim = 512 cap = 512]
```

第 11 章 文字编解码

CHAPTER 11

文字编解码定义字符集、解码器、编码器。用于在字节和 Unicode 字符之间进行转换。

11.1 Charset 字符集

Unicode 是代码单元序列与字节序列之间的命名映射。此类定义用于创建解码器和编码器及用于检索与字符集相关联的各种名称的方法，此类的实例不可改变并发安全。

本节基于官方文档介绍常用方法，核心方法配合代码示例以方便读者理解。

1. availableCharsets()

静态方法，返回从规范字符集名称到字符集对象的已排序集合，代码如下：

```
//第 11 章/one/CharsetTest.java
public class CharsetTest {

    public static void main(String[] args) {
        SortedMap<String, Charset> charsets = Charset.availableCharsets();
        System.out.println(charsets);
    }
}
```

执行结果如下：

```
{Big5 = Big5, Big5 - HKSCS = Big5 - HKSCS, CESU - 8 = CESU - 8, EUC - JP = EUC - JP, EUC - KR = EUC -
KR, GB18030 = GB18030, GB2312 = GB2312, GBK = GBK, IBM - Thai = IBM - Thai, IBM00858 = IBM00858,
IBM01140 = IBM01140, IBM01141 = IBM01141, IBM01142 = IBM01142, IBM01143 = IBM01143, IBM01144
= IBM01144, IBM01145 = IBM01145, IBM01146 = IBM01146, IBM01147 = IBM01147, IBM01148 =
IBM01148, IBM01149 = IBM01149, IBM037 = IBM037, IBM1026 = IBM1026, IBM1047 = IBM1047, IBM273 =
IBM273, IBM277 = IBM277, IBM278 = IBM278, IBM280 = IBM280, IBM284 = IBM284, IBM285 = IBM285,
IBM290 = IBM290, IBM297 = IBM297, IBM420 = IBM420, IBM424 = IBM424, IBM437 = IBM437, IBM500 =
IBM500, IBM775 = IBM775, IBM850 = IBM850, IBM852 = IBM852, IBM855 = IBM855, IBM857 = IBM857,
IBM860 = IBM860, IBM861 = IBM861, IBM862 = IBM862, IBM863 = IBM863, IBM864 = IBM864, IBM865 =
IBM865, IBM866 = IBM866, IBM868 = IBM868, IBM869 = IBM869, IBM870 = IBM870, IBM871 = IBM871,
IBM918 = IBM918, ISO - 2022 - CN = ISO - 2022 - CN, ISO - 2022 - JP = ISO - 2022 - JP, ISO - 2022 - JP
- 2 = ISO - 2022 - JP - 2, ISO - 2022 - KR = ISO - 2022 - KR, ISO - 8859 - 1 = ISO - 8859 - 1, ISO -
8859 - 13 = ISO - 8859 - 13, ISO - 8859 - 15 = ISO - 8859 - 15, ISO - 8859 - 16 = ISO - 8859 - 16, ISO -
```

```
8859 - 2 = ISO - 8859 - 2, ISO - 8859 - 3 = ISO - 8859 - 3, ISO - 8859 - 4 = ISO - 8859 - 4, ISO - 8859
 - 5 = ISO - 8859 - 5, ISO - 8859 - 6 = ISO - 8859 - 6, ISO - 8859 - 7 = ISO - 8859 - 7, ISO - 8859 - 8
 = ISO - 8859 - 8, ISO - 8859 - 9 = ISO - 8859 - 9, JIS_X0201 = JIS_X0201, JIS_X0212 - 1990 = JIS_
X0212 - 1990, KOI8 - R = KOI8 - R, KOI8 - U = KOI8 - U, Shift_JIS = Shift_JIS, TIS - 620 = TIS -
620, US - ASCII = US - ASCII, UTF - 16 = UTF - 16, UTF - 16BE = UTF - 16BE, UTF - 16LE = UTF - 16LE,
UTF - 32 = UTF - 32, UTF - 32BE = UTF - 32BE, UTF - 32LE = UTF - 32LE, UTF - 8 = UTF - 8, windows -
1250 = windows - 1250, windows - 1251 = windows - 1251, windows - 1252 = windows - 1252, windows
 - 1253 = windows - 1253, windows - 1254 = windows - 1254, windows - 1255 = windows - 1255,
windows - 1256 = windows - 1256, windows - 1257 = windows - 1257, windows - 1258 = windows - 1258,
windows - 31j = windows - 31j, x - Big5 - HKSCS - 2001 = x - Big5 - HKSCS - 2001, x - Big5 - Solaris
 = x - Big5 - Solaris, x - euc - jp - linux = x - euc - jp - linux, x - EUC - TW = x - EUC - TW, x -
eucJP - Open = x - eucJP - Open, x - IBM1006 = x - IBM1006, x - IBM1025 = x - IBM1025, x - IBM1046 =
x - IBM1046, x - IBM1097 = x - IBM1097, x - IBM1098 = x - IBM1098, x - IBM1112 = x - IBM1112, x -
IBM1122 = x - IBM1122, x - IBM1123 = x - IBM1123, x - IBM1124 = x - IBM1124, x - IBM1129 = x -
IBM1129, x - IBM1166 = x - IBM1166, x - IBM1364 = x - IBM1364, x - IBM1381 = x - IBM1381, x -
IBM1383 = x - IBM1383, x - IBM29626C = x - IBM29626C, x - IBM300 = x - IBM300, x - IBM33722 = x -
IBM33722, x - IBM737 = x - IBM737, x - IBM833 = x - IBM833, x - IBM834 = x - IBM834, x - IBM856 = x
 - IBM856, x - IBM874 = x - IBM874, x - IBM875 = x - IBM875, x - IBM921 = x - IBM921, x - IBM922 = x
 - IBM922, x - IBM930 = x - IBM930, x - IBM933 = x - IBM933, x - IBM935 = x - IBM935, x - IBM937 = x
 - IBM937, x - IBM939 = x - IBM939, x - IBM942 = x - IBM942, x - IBM942C = x - IBM942C, x - IBM943 =
x - IBM943, x - IBM943C = x - IBM943C, x - IBM948 = x - IBM948, x - IBM949 = x - IBM949, x - IBM949C
 = x - IBM949C, x - IBM950 = x - IBM950, x - IBM964 = x - IBM964, x - IBM970 = x - IBM970, x -
ISCII91 = x - ISCII91, x - ISO - 2022 - CN - CNS = x - ISO - 2022 - CN - CNS, x - ISO - 2022 - CN - GB
 = x - ISO - 2022 - CN - GB, x - iso - 8859 - 11 = x - iso - 8859 - 11, x - JIS0208 = x - JIS0208, x -
JISAutoDetect = x - JISAutoDetect, x - Johab = x - Johab, x - MacArabic = x - MacArabic, x -
MacCentralEurope = x - MacCentralEurope, x - MacCroatian = x - MacCroatian, x - MacCyrillic = x -
MacCyrillic, x - MacDingbat = x - MacDingbat, x - MacGreek = x - MacGreek, x - MacHebrew = x -
MacHebrew, x - MacIceland = x - MacIceland, x - MacRoman = x - MacRoman, x - MacRomania = x -
MacRomania, x - MacSymbol = x - MacSymbol, x - MacThai = x - MacThai, x - MacTurkish = x -
MacTurkish, x - MacUkraine = x - MacUkraine, x - MS932_0213 = x - MS932_0213, x - MS950 - HKSCS =
x - MS950 - HKSCS, x - MS950 - HKSCS - XP = x - MS950 - HKSCS - XP, x - mswin - 936 = x - mswin - 936,
x - PCK = x - PCK, x - SJIS_0213 = x - SJIS_0213, x - UTF - 16LE - BOM = x - UTF - 16LE - BOM, X - UTF
 - 32BE - BOM = X - UTF - 32BE - BOM, X - UTF - 32LE - BOM = X - UTF - 32LE - BOM, x - windows - 50220 =
x - windows - 50220, x - windows - 50221 = x - windows - 50221, x - windows - 874 = x - windows -
874, x - windows - 949 = x - windows - 949, x - windows - 950 = x - windows - 950, x - windows -
iso2022jp = x - windows - iso2022jp}
```

2. isSupported(String charsetName)

静态方法，检查是否支持指定的字符集名称，返回 boolean 值。接收 String 入参，作为指定的字符集名称，代码如下：

```java
//第 11 章/one/CharsetTest.java
public class CharsetTest {

   public static void main(String[] args) {
      System.out.println(Charset.isSupported("UTF - 8"));
   }
}
```

执行结果如下：

```
true
```

3. forName(String charsetName)

静态方法，返回指定名称的字符集对象。接收 String 入参，作为指定的字符集名称，代码如下：

```
//第 11 章/one/CharsetTest.java
public class CharsetTest {

    public static void main(String[] args) {
        Charset charset = Charset.forName("UTF-8");
        System.out.println(charset);
    }
}
```

执行结果如下：

```
UTF-8
```

4. aliases()

返回当前字符集的别名集合，代码如下：

```
//第 11 章/one/CharsetTest.java
public class CharsetTest {

    public static void main(String[] args) {
        Charset charset = StandardCharsets.UTF_8;
        System.out.println(charset);
        System.out.println(charset.aliases());
    }
}
```

执行结果如下：

```
UTF-8
[unicode-1-1-utf-8, UTF8]
```

5. canEncode()

检查当前字符集是否支持编码，返回 boolean 值。

6. compareTo(Charset that)

比较当前字符集和指定的字符集是否相同，返回 int 值。接收 Charset 入参，作为指定的字符集。

7. contains(Charset cs)

检查当前字符集是否包含指定的字符集，返回 boolean 值。接收 Charset 入参，作为指定的字符集。

8. decode(ByteBuffer bb)

将指定的字节数据解码为字符数据。接收 ByteBuffer 入参,作为指定的字节数据,代码如下:

```
//第 11 章/one/CharsetTest.java
public class CharsetTest {

    public static void main(String[] args) {
        Charset charset = StandardCharsets.UTF_8;
        ByteBuffer byteBuffer = ByteBuffer.allocate(64);
        byteBuffer.put((byte) 65).put((byte) 66).put((byte) 67);
        byteBuffer.flip();
        CharBuffer decode = charset.decode(byteBuffer);
        System.out.println(decode);
        System.out.println(charset.displayName());
        System.out.println(charset.displayName(Locale.CHINESE));
        String s = new String(new byte[]{65, 66, 67},
                                        StandardCharsets.ISO_8859_1);
        System.out.println(s);
        byte[] bytes = s.getBytes(StandardCharsets.ISO_8859_1);
        System.out.println(Arrays.toString(bytes));
    }
}
```

执行结果如下:

```
ABC
UTF-8
UTF-8
ABC
[65, 66, 67]
```

9. defaultCharset()

静态方法,返回当前环境默认的字符集对象。

10. displayName()

返回默认语言环境下当前字符集的可读名称。

11. displayName(Locale locale)

返回指定语言环境下当前字符集的可读名称。接收 Locale 入参,作为指定语言环境。

12. encode(String str)

将指定的字符串数据编码为字节数据。接收 String 入参,作为指定的字符串数据。

13. encode(CharBuffer cb)

将指定的字符数据编码为字节数据。接收 CharBuffer 入参,作为指定的字符数据。

14. equals(Object ob)

比较当前对象和指定的对象，返回 boolean 值。接收 Object 入参，作为指定的对象。

15. isRegistered()

检查当前字符集是否在 IANA 注册表中注册，返回 boolean 值。

16. name()

返回当前字符集的可读名称。

17. newDecoder()

从当前字符集构造新的解码器。

18. newEncoder()

从当前字符集构造新的编码器。

11.2 CharsetEncoder 编码器

CharsetEncoder 编码器是一种引擎，它可以将 Unicode 字符序列转换为指定字符集中的字节序列。本节基于官方文档介绍常用方法，核心方法配合代码示例以方便读者理解。

1. averageBytesPerChar()

返回将为每个输入字符生成的平均字节数，代码如下：

```
//第 11 章/two/CharsetEncoderTest.java
public class CharsetEncoderTest {

    public static void main(String[] args) throws Exception {
        Charset charset = StandardCharsets.UTF_8;
        CharsetEncoder encoderUTF8 = charset.newEncoder();
        System.out.println(encoderUTF8.averageBytesPerChar());
        System.out.println(encoderUTF8.maxBytesPerChar());

        Charset charsetGBK = Charset.forName("GBK");
        CharsetEncoder encodeGBK = charsetGBK.newEncoder();
        System.out.println(encodeGBK.averageBytesPerChar());
        System.out.println(encodeGBK.maxBytesPerChar());
    }
}
```

执行结果如下：

```
1.1
3.0
2.0
2.0
```

2. maxBytesPerChar()

返回将为每个输入字符生成的最大字节数。

3. canEncode(char c)

检查当前编码器是否可以对指定的字符进行编码,返回 boolean 值。接收 char 入参,作为指定的字符。

4. canEncode(CharSequence cs)

检查当前编码器是否可以对指定的字符序列进行编码,返回 boolean 值。接收 CharSequence 入参,作为指定的字符序列。

5. charset()

返回当前编码器的字符集对象。

6. encode(CharBuffer in)

对指定的字符缓冲区的内容进行编码并存储在新分配的字节缓冲区。接收 CharBuffer 入参,作为指定的字符缓冲区,代码如下:

```
//第 11 章/two/CharsetEncoderTest.java
public class CharsetEncoderTest {

    public static void main(String[] args) throws Exception {
        //UTF-8
        Charset charset = StandardCharsets.UTF_8;
        CharsetEncoder charsetUTF8 = charset.newEncoder();
        System.out.println(charsetUTF8.canEncode('我'));
        System.out.println(charsetUTF8.canEncode('A'));
        CharBuffer charBuffer = CharBuffer.allocate(1024);
        charBuffer.put("你好世界-helloworld");
        charBuffer.flip();
        ByteBuffer encode = charsetUTF8.encode(charBuffer);
        System.out.println(encode);
        //GBK
        Charset charsetGBK = Charset.forName("GBK");
        CharsetEncoder encoderGBK = charsetGBK.newEncoder();
        CharBuffer charBufferGBK = CharBuffer.allocate(1024);
        charBufferGBK.put("你好世界-helloworld");
        charBufferGBK.flip();
        ByteBuffer bufferGBK = encoderGBK.encode(charBufferGBK);
        System.out.println(bufferGBK);
    }
}
```

执行结果如下:

```
true
true
```

```
java.nio.HeapByteBuffer[pos = 0 lim = 23 cap = 33]
java.nio.HeapByteBuffer[pos = 0 lim = 19 cap = 30]
```

7. encode(CharBuffer in, ByteBuffer out, boolean endOfInput)

对指定的字符缓冲区的内容进行编码并存储在指定的字节缓冲区。接收 CharBuffer 入参，作为指定的字符缓冲区，接收 ByteBuffer 入参，作为指定的字节缓冲区，接收 boolean 入参，作为是否结束输入的标记，代码如下：

```
//第 11 章/two/CharsetEncoderTest.java
public class CharsetEncoderTest {

    public static void main(String[] args) throws Exception {
        Charset charset = StandardCharsets.UTF_8;
        CharsetEncoder charsetEncoder = charset.newEncoder();
        CharBuffer charBuffer = CharBuffer.allocate(1024);
        charBuffer.put("你好世界 - helloworld");
        charBuffer.flip();
        ByteBuffer byteBuffer = ByteBuffer.allocate(1024);
        CoderResult coderResult = charsetEncoder.encode(charBuffer,
                                             byteBuffer, true);
        if(coderResult.isUnderflow()){
            coderResult = charsetEncoder.flush(byteBuffer);
        }
        System.out.println(coderResult.isError());
        System.out.println(coderResult);
        System.out.println(byteBuffer);
    }
}
```

执行结果如下：

```
false
UNDERFLOW
java.nio.HeapByteBuffer[pos = 23 lim = 1024 cap = 1024]
```

8. isLegalReplacement(byte[] repl)

检查指定的字节数组是否为当前编码器的合法替换值，返回 boolean 值。接收 byte[] 入参，作为指定的字节数组。

9. malformedInputAction()

返回当前输入格式错误的操作对象，代码如下：

```
//第 11 章/two/CharsetEncoderTest.java
public class CharsetEncoderTest {

    public static void main(String[] args) throws Exception {
        Charset charset = StandardCharsets.UTF_8;
        CharsetEncoder charsetEncoder = charset.newEncoder();
```

```
        System.out.println(charsetEncoder.malformedInputAction());
        System.out.println(charsetEncoder.unmappableCharacterAction());
    }
}
```

执行结果如下：

```
REPORT
REPORT
```

10. unmappableCharacterAction()

返回当前不可映射字符错误的操作对象。

11. onMalformedInput(CodingErrorAction newAction)

设置当前输入格式错误的操作对象。接收 CodingErrorAction 入参，作为输入格式错误的操作对象。

12. onUnmappableCharacter(CodingErrorAction newAction)

设置当前不可映射字符错误的操作对象。接收 CodingErrorAction 入参，作为不可映射字符错误的操作对象。

13. replacement()

返回当前编码器的替换值，代码如下：

```
//第 11 章/two/CharsetEncoderTest.java
public class CharsetEncoderTest {

    public static void main(String[] args) throws Exception {
        Charset charset = StandardCharsets.UTF_8;
        CharsetEncoder charsetEncoder = charset.newEncoder();
        System.out.println(Arrays.toString(charsetEncoder.replacement()));
    }
}
```

执行结果如下：

```
[63]
```

14. replaceWith(byte[] newReplacement)

设置当前编码器的替换值。接收 byte[]入参，作为编码器的替换值。

15. reset()

重置当前编码器并清除任何内部状态。

11.3 CoderResult 类

编码器、解码器的结果状态描述，核心状态见表 11-1。

表 11-1　编码器、解码器的核心状态

状　　态	描　　述
UNDERFLOW	当没有更多的输入数据时
OVERFLOW	当输出缓冲区的剩余空间不足时
MALFORMED	当输入单元序列格式不正确时
UNMAPPABLE	当输入单元序列报告不可映射字符错误时

本节基于官方文档介绍常用方法，核心方法配合代码示例以方便读者理解。

1. malformedForLength(int length)

静态方法，创建描述输入单元序列格式不正确时的对象，代码如下：

```
//第 11 章/three/CoderResultTest.java
public class CoderResultTest {

    public static void main(String[] args) {
        CoderResult coderResult = CoderResult.malformedForLength(2);
        System.out.println(coderResult.isError());
        System.out.println(coderResult.isMalformed());
        System.out.println(coderResult.isUnmappable());
        System.out.println(coderResult.length());
        System.out.println(coderResult);
    }
}
```

执行结果如下：

```
true
true
false
2
MALFORMED[2]
```

2. unmappableForLength(int length)

静态方法，创建描述输入单元序列报告不可映射字符错误时的对象，代码如下：

```
//第 11 章/three/CoderResultTest.java
public class CoderResultTest {

    public static void main(String[] args) {
        CoderResult coderResult = CoderResult.unmappableForLength(2);
        System.out.println(coderResult.isError());
        System.out.println(coderResult.isMalformed());
        System.out.println(coderResult.isUnmappable());
        System.out.println(coderResult.length());
        System.out.println(coderResult);
    }
}
```

执行结果如下：

```
true
false
true
2
UNMAPPABLE[2]
```

3. isError()

检查当前对象是否描述了错误，返回 boolean 值。

4. isMalformed()

检查当前对象是否描述了输入单元序列格式不正确的错误，返回 boolean 值。

5. isUnmappable()

检查当前对象是否描述了输入单元序列报告不可映射字符的错误，返回 boolean 值。

6. isOverflow()

检查当前对象是否描述了没有更多的输入数据，返回 boolean 值。

7. isUnderflow()

检查当前对象是否描述了输出缓冲区的剩余空间不足，返回 boolean 值。

8. throwException()

抛出当前对象描述的异常信息。

11.4 CharsetDecoder 解码器

CharsetDecoder 解码器是一种引擎，可以将字节序列转换为指定字符集中的 Unicode 字符序列。本节基于官方文档介绍常用方法，核心方法配合代码示例以方便读者理解。

1. averageCharsPerByte()

返回将为每个输入字节生成的平均字符数，代码如下：

```
//第 11 章/four/CharsetDecoderTest.java
public class CharsetDecoderTest {

    public static void main(String[] args) throws Exception {
        Charset charset = StandardCharsets.UTF_8;
        CharsetDecoder charsetDecoder = charset.newDecoder();
        System.out.println(charsetDecoder.averageCharsPerByte());
        System.out.println(charsetDecoder.maxCharsPerByte());
    }
}
```

执行结果如下：

```
1.0
1.0
```

2. maxCharsPerByte()

返回将为每个输入字节生成的最大字符数。

3. charset()

返回当前解码器的字符集对象。

4. decode(ByteBuffer in)

对指定的字节缓冲区的内容进行解码并存储在新分配的字符缓冲区。接收 ByteBuffer 入参，作为指定的字节缓冲区，代码如下：

```
//第 11 章/four/CharsetDecoderTest.java
public class CharsetDecoderTest {

    public static void main(String[] args) throws Exception {
        Charset charset = StandardCharsets.UTF_8;
        CharsetDecoder charsetDecoder = charset.newDecoder();
        ByteBuffer byteBuffer = ByteBuffer.allocate(1024);
        byteBuffer.put("helloWorld - 你好世界".
                                    getBytes(StandardCharsets.UTF_8));
        byteBuffer.flip();
        CharBuffer decode = charsetDecoder.decode(byteBuffer);
        System.out.println(decode);
    }
}
```

执行结果如下：

```
helloWorld - 你好世界
```

5. decode(ByteBuffer in, CharBuffer out, boolean endOfInput)

对指定的字节缓冲区的内容进行解码并存储在指定的字符缓冲区。接收 ByteBuffer 入参，作为指定的字节缓冲区，接收 CharBuffer 入参，作为指定的字符缓冲区，接收 boolean 入参，作为是否结束输入的标记，代码如下：

```
//第 11 章/four/CharsetDecoderTest.java
public class CharsetDecoderTest {

    public static void main(String[] args) throws Exception {
        Charset charset = StandardCharsets.UTF_8;
        CharsetDecoder charsetDecoder = charset.newDecoder();
        ByteBuffer byteBuffer = ByteBuffer.allocate(1024);
        byteBuffer.put("helloWorld - 你好世界".
                                   getBytes(StandardCharsets.UTF_8));
        byteBuffer.flip();
        byteBuffer.limit(byteBuffer.limit() - 1);
```

```
        CharBuffer charBuffer = CharBuffer.allocate(1024);
        CoderResult result = charsetDecoder.
                                    decode(byteBuffer,charBuffer,false);
        charBuffer.flip();
        System.out.println(result);
        System.out.println(byteBuffer);
        System.out.println(charBuffer);
        byteBuffer.limit(byteBuffer.limit() + 1);
        charBuffer.clear();
        CoderResult resultEnd = charsetDecoder.
                                    decode(byteBuffer, charBuffer, true);
        charBuffer.flip();
        System.out.println(resultEnd);
        System.out.println(byteBuffer);
        System.out.println(charBuffer);
    }
}
```

执行结果如下：

```
UNDERFLOW
java.nio.HeapByteBuffer[pos = 20 lim = 22 cap = 1024]
helloWorld - 你好世
UNDERFLOW
java.nio.HeapByteBuffer[pos = 23 lim = 23 cap = 1024]
界
```

6. detectedCharset()

返回当前解码器检测到的字符集对象。此方法默认抛出异常，如图 11-1 所示。

```
public Charset detectedCharset() {
    throw new UnsupportedOperationException();
}
```

图 11-1　方法源代码

7. isAutoDetecting()

检查当前解码器是否自动检测字符集，返回 boolean 值，代码如下：

```
//第 11 章/four/CharsetDecoderTest.java
public class CharsetDecoderTest {

    public static void main(String[] args) throws Exception {
        Charset charset = StandardCharsets.UTF_8;
        CharsetDecoder charsetDecoder = charset.newDecoder();
        System.out.println(charsetDecoder.isAutoDetecting());
    }
}
```

执行结果如下：

```
false
```

8. isCharsetDetected()

检查当前解码器是否尚未检测到字符集，返回 boolean 值。此方法默认抛出异常，如

```
public boolean isCharsetDetected() {
    throw new UnsupportedOperationException();
}
```

图 11-2 方法源代码

图 11-2 所示。

9. malformedInputAction()

返回当前输入格式错误的操作对象。

10. unmappableCharacterAction()

返回当前不可映射字符错误的操作对象。

11. onMalformedInput(CodingErrorAction newAction)

设置当前输入格式错误的操作对象。接收 CodingErrorAction 入参，作为输入格式错误的操作对象。

12. onUnmappableCharacter(CodingErrorAction newAction)

设置当前不可映射字符错误的操作对象。接收 CodingErrorAction 入参，作为不可映射字符错误的操作对象。

13. replacement()

返回当前解码器的替换值，代码如下：

```
//第 11 章/four/CharsetDecoderTest.java
public class CharsetDecoderTest {

    public static void main(String[] args) throws Exception {
        Charset charset = StandardCharsets.UTF_8;
        CharsetDecoder charsetDecoder = charset.newDecoder();
        System.out.println(charsetDecoder.replacement());
    }
}
```

执行结果如下：

14. replaceWith(String newReplacement)

设置当前解码器的替换值。接收 String 入参，作为解码器的替换值。

15. reset()

重置当前解码器并清除任何内部状态。

小结

本章介绍了 Unicode 字符、字符集、字符集编码器、字符集解码器。需要理解 Unicode 字符和字符集的区别。

习题

1. 判断题

(1) Unicode 字符是规范的、通用的文字表示系统。( )

(2) 编码器可以将 Unicode 字符序列转换为指定字符集中的字节序列。( )

(3) 解码器可以将字节序列转换为指定字符集中的 Unicode 字符序列。( )

2. 选择题

(1) 规范的字符集有( )。(多选)

A. GBK　　B. UTF-8

C. ISO-8859-1　　D. UTF-9

(2) 能表示汉字的字符集有( )。(多选)

A. GBK　　B. UTF-8

C. ISO-8859-1　　D. UTF-9

(3) 表示字符集编码器的类是( )。(单选)

A. CoderResult 类　　B. CharsetEncoder 类

C. CharsetDecoder 类　　D. CodingErrorAction 类

(4) 表示字符集解码器的类是( )。(单选)

A. CoderResult 类　　B. CharsetEncoder 类

C. CharsetDecoder 类　　D. CodingErrorAction 类

(5) 表示字符集编解码结果的类是( )。(单选)

A. CoderResult 类　　B. CharsetEncoder 类

C. CharsetDecoder 类　　D. CodingErrorAction 类

3. 填空题

(1) 查看执行结果并补充代码,代码如下:

```
//第 11 章/answer/OneAnswer.java
public class OneAnswer {

    public static void main(String[] args) {
        Charset charset = StandardCharsets.UTF_8;
        CharsetDecoder charsetDecoder = charset.newDecoder();
        ByteBuffer byteBuffer = ByteBuffer.allocate(1024);
        byteBuffer.put("helloWorld - 你好世界"
                            .getBytes(StandardCharsets.UTF_8));
        byteBuffer.flip();
        byteBuffer.limit(________________);
        CharBuffer charBuffer = CharBuffer.allocate(1024);
        CoderResult result = charsetDecoder
```

```
                                .decode(byteBuffer,charBuffer,false);
        charBuffer.flip();
        System.out.println(result);
        System.out.println(byteBuffer);
        System.out.println(charBuffer);
        byteBuffer.limit(________________);
        charBuffer.clear();
        CoderResult resultEnd = charsetDecoder
                                .decode(byteBuffer, charBuffer, true);
        charBuffer.flip();
        System.out.println(resultEnd);
        System.out.println(byteBuffer);
        System.out.println(charBuffer);
    }
}
```

执行结果如下：

```
UNDERFLOW
java.nio.HeapByteBuffer[pos = 20 lim = 22 cap = 1024]
helloWorld - 你好世
UNDERFLOW
java.nio.HeapByteBuffer[pos = 23 lim = 23 cap = 1024]
界
```

（2）查看执行结果并补充代码，代码如下：

```
//第 11 章/answer/TwoAnswer.java
public class TwoAnswer {

    public static void main(String[] args) {
        Charset charset = StandardCharsets.UTF_8;
        ByteBuffer byteBuffer = ByteBuffer.allocate(64);
        byteBuffer.put(________).put(________).put(________);
        byteBuffer.flip();
        CharBuffer decode = charset.decode(byteBuffer);
        System.out.println(decode);
        System.out.println(charset.displayName());
        String s = new String(new byte[]{________, ________, ________},
                                StandardCharsets.ISO_8859_1);
        System.out.println(s);
        byte[] bytes = s.getBytes(StandardCharsets.ISO_8859_1);
        System.out.println(Arrays.toString(bytes));
    }
}
```

执行结果如下：

```
ABC
UTF - 8
ABC
[65, 66, 67]
```

第 12 章 网络通道

CHAPTER 12

channels 定义通道，表示能够执行 I/O 操作（文件和套接字）的实体连接。还可定义选择器，用于多路复用的非阻塞式 I/O 操作。

12.1 FileChannel 抽象类

FileChannel 抽象类表示用于读取、写入、映射、操作文件的通道。本节基于官方文档介绍常用方法，核心方法配合代码示例以方便读者理解。

1. open(Path path, OpenOption... options)

静态方法，打开指定的文件并返回表示此文件通道的对象。接收 Path 入参，作为文件的路径，接收 OpenOption 入参，作为文件的选项，代码如下：

```
//第 12 章/one/FileChannelTest.java
public class FileChannelTest {

    public static void main(String[] args) {
        try (FileChannel fileChannel = FileChannel.
                                open(Path.of("D:\\temp\\src\\test5.txt"),
                StandardOpenOption.READ, StandardOpenOption.WRITE)) {
            System.out.println(fileChannel.position());
            System.out.println(fileChannel.isOpen());
        } catch (Exception e) {
            e.printStackTrace();
        }
    }
}
```

执行结果如下：

```
0
true
```

2. open(Path path, Set <? extends OpenOption > options, FileAttribute <?>... attrs)

静态方法，打开指定的文件并返回表示此文件通道的对象。接收 Path 入参，作为文件的路径，接收 OpenOption 入参，作为文件的选项，接收 FileAttribute 入参，作为文件属性的值。

3. force(boolean metaData)

强制将当前文件通道的任何更新同步到底层的存储设备。接收 boolean 入参，如果为真，则强制将文件的内容和元数据同步到存储设备，否则只需强制同步文件的内容。

4. lock()

获取当前文件通道的互斥文件锁对象，代码如下：

```
//第 12 章/one/FileChannelTest.java
public class FileChannelTest {

    public static void main(String[] args) {
        try (FileChannel fileChannel = FileChannel.
                                open(Path.of("D:\\temp\\src\\test5.txt"),
                StandardOpenOption.READ, StandardOpenOption.WRITE);
             FileLock lock = fileChannel.lock()) {
            System.out.println(fileChannel.position());
            System.out.println(fileChannel.size());
            System.out.println(lock.isValid());
            System.out.println(lock.isShared());
        } catch (Exception e) {
            e.printStackTrace();
        }
    }
}
```

执行结果如下：

```
0
510
true
false
```

此方法的源代码如图 12-1 所示。

```
public final FileLock lock() throws IOException {
    return lock(0L, Long.MAX_VALUE, false);
}
```

图 12-1　源代码

5. lock(long position, long size, boolean shared)

获取当前文件通道的文件锁对象。接收 long 入参，作为锁定区域的开始位置，接收 long 入参，作为锁定区域的大小，接收 boolean 入参，作为是否共享锁定。

6. map(FileChannel.MapMode mode, long position, long size)

将当前文件通道指定区域的内容直接映射到内存中。接收 FileChannel.MapMode 入参，作为文件的映射方式，接收 long 入参，作为映射区域的开始位置，接收 long 入参，作为映射区域的大小。

7. position()

返回当前文件通道数据指针的位置。

8. position(long newPosition)

设置当前文件通道数据指针的位置。接收 long 入参,作为数据指针的位置,代码如下:

```java
//第 12 章/one/FileChannelTest.java
public class FileChannelTest {

    public static void main(String[] args) {
        try (FileChannel fileChannel = FileChannel.
                                open(Path.of("D:\\temp\\src\\test5.txt"),
                StandardOpenOption.READ, StandardOpenOption.WRITE)) {
            System.out.println(fileChannel.position());
            System.out.println(fileChannel.size());
            ByteBuffer byteBuffer = ByteBuffer.allocate(10);
            System.out.println(fileChannel.read(byteBuffer));
            byteBuffer.flip();
            System.out.println(new String(byteBuffer.array(),
                        byteBuffer.position(), byteBuffer.limit()));
            System.out.println(fileChannel.position());
            fileChannel.position(0);
            System.out.println(fileChannel.position());
        } catch (Exception e) {
            e.printStackTrace();
        }
    }
}
```

执行结果如下:

```
0
510
10
helloworld
10
0
```

9. read(ByteBuffer dst)

读取当前文件通道的数据并存储到指定的字节缓冲区,返回读取到的数据长度。接收 ByteBuffer 入参,作为指定的字节缓冲区。

10. read(ByteBuffer dst, long position)

读取当前文件通道的数据并存储到指定的字节缓冲区,返回读取到的数据长度。接收 ByteBuffer 入参,作为指定的字节缓冲区,接收 long 入参,作为数据指针的开始位置。

11. size()

返回当前文件通道的当前数据大小,以字节为单位。

12. truncate(long size)

将当前文件通道的数据截断为指定的大小。接收 long 入参，作为指定的大小。

13. tryLock()

尝试获取当前文件通道的互斥文件锁对象。

14. tryLock(long position, long size, boolean shared)

尝试获取当前文件通道的文件锁对象。接收 long 入参，作为锁定区域的开始位置，接收 long 入参，作为锁定区域的大小，接收 boolean 入参，作为是否共享锁定。

15. write(ByteBuffer src)

写入当前文件通道的数据，返回写入的数据长度。接收 ByteBuffer 入参，作为指定的字节缓冲区，代码如下：

```
//第 12 章/one/FileChannelTest.java
public class FileChannelTest {

    public static void main(String[] args) {
        try(FileChannel fileChannel = FileChannel.
                                open(Path.of("D:\\temp\\src\\test5.txt"),
                StandardOpenOption.READ, StandardOpenOption.WRITE);
            FileLock lock = fileChannel.lock(0,fileChannel.size(),false)) {
            System.out.println(fileChannel.position());
            System.out.println(fileChannel.size());
            System.out.println(lock.isValid());

            ByteBuffer byteBuffer = ByteBuffer.allocate(10);
            System.out.println(fileChannel.read(byteBuffer));
            byteBuffer.flip();
            System.out.println(new String(byteBuffer.array(),
                        byteBuffer.position(), byteBuffer.limit()));
            System.out.println(fileChannel.position());

            fileChannel.position(0);

            byteBuffer.clear();
            byteBuffer.put("helloworld".getBytes());
            byteBuffer.flip();
            System.out.println(fileChannel.write(byteBuffer));
        } catch (Exception e) {
            e.printStackTrace();
        }
    }
}
```

执行结果如下：

```
0
510
true
10
helloworld
10
10
```

16. write(ByteBuffer src, long position)

写入当前文件通道的数据,返回写入的数据长度。接收 ByteBuffer 入参,作为指定的字节缓冲区,接收 long 入参,作为数据指针的开始位置。

17. close()

关闭当前文件通道并释放资源。

18. isOpen()

检查当前文件通道是否已打开,返回 boolean 值。

12.2 StandardOpenOption 枚举类

StandardOpenOption 枚举类用于定义标准的打开选项枚举对象,枚举对象见表 12-1。

表 12-1　枚举对象

枚举对象	描　述
APPEND	如果是 WRITE 访问打开了文件,则字节将写入文件末尾而不是开头
CREATE	创建新文件,如果该文件不存在
CREATE_NEW	创建新文件,如果该文件已存在,则将失败
DELETE_ON_CLOSE	关闭时删除该文件
DSYNC	要求将文件内容的每个更新同步到底层存储设备
READ	文件的读取模式
SPARSE	稀疏文件
SYNC	要求将文件内容或元数据的每个更新同步到底层存储设备
TRUNCATE_EXISTING	如果该文件已存在并且它是 WRITE 访问打开的,则它的长度将被截断
WRITE	文件的写入模式

12.3 FileLock 抽象类

FileLock 抽象类表示锁定的文件区域对象。本节基于官方文档介绍常用方法,核心方法配合代码示例以方便读者理解。

1. acquiredBy()

返回当前锁定文件区域对象的通道。

2. channel()

返回当前锁定文件区域对象的文件通道。

3. close()

关闭当前锁定文件区域对象并释放资源。

4. isShared()

检查当前锁定文件区域对象是否是共享模式，返回 boolean 值。

5. isValid()

检查当前锁定文件区域对象是否有效，返回 boolean 值。

6. overlaps(long position, long size)

检查当前锁定文件区域对象是否重叠，返回 boolean 值。接收 long 入参，作为锁定区域的开始位置，接收 long 入参，作为锁定区域的大小。

7. position()

返回锁定文件区域对象数据指针的位置。

8. release()

关闭当前锁定文件区域对象并释放资源。

9. size()

返回当前锁定文件区域对象的数据大小，以字节为单位。

42min

12.4 ServerSocketChannel 抽象类

此类表示服务器端套接字通道，通过选择器监听客户端套接字请求并通过网络连接，当网络连接完成后便可以进行双向通信。

12.4.1 常用方法

本节基于官方文档介绍常用方法，核心方法配合代码示例以方便读者理解。

1. bind(SocketAddress local)

将当前套接字绑定到指定的套接字地址。接收 SocketAddress 入参，作为指定的套接字地址。

2. bind(SocketAddress local, int backlog)

将当前套接字绑定到指定的套接字地址。接收 SocketAddress 入参,作为指定的套接字地址,接收 int 入参,作为指定连接的最大队列长度。

3. getLocalAddress()

返回当前套接字所绑定的套接字地址。

4. setOption(SocketOption<T> name, T value)

设置当前套接字指定选项的值。接收 SocketOption 入参,作为指定选项,接收 T 入参,作为指定选项的值。

5. socket()

返回当前的服务器端套接字对象。

6. validOps()

返回当前通道受支持操作的操作集标识。

7. configureBlocking(boolean block)

设置当前套接字的阻塞方式。接收 boolean 入参,作为套接字的阻塞方式。

8. isBlocking()

检查当前套接字的阻塞方式,返回 boolean 值。

9. isRegistered()

检查当前通道是否向任何选择器注册,返回 boolean 值。

10. keyFor(Selector sel)

返回指定的选择器检索到的当前通道注册的键。接收 Selector 入参,作为指定的选择器。

11. provider()

返回创建当前服务器端套接字通道的提供者。

12. register(Selector sel, int ops, Object att)

把当前通道注册到指定的选择器中。接收 Selector 入参,作为指定的选择器,接收 int 入参,作为操作集标识,接收 Object 入参,作为附加的数据。返回 SelectionKey 对象。

13. close()

关闭当前套接字通道并释放资源。

14. isOpen()

检查当前通道是否已打开,返回 boolean 值。

15. getOption(SocketOption<T> name)

返回当前套接字指定选项的值。接收 SocketOption 入参,作为指定选项。

16. supportedOptions()

返回当前套接字所支持的套接字选项集合。

12.4.2 使用示例

使用非阻塞模式设计一个服务器端，接收客户端的连接，连接完成后向客户端发送固定数据。数据发送完毕后关闭此次客户端连接，代码如下：

```
//第 12 章/four/ServerSocketChannelTest.java
public class ServerSocketChannelTest {

    private static final ByteBuffer BUFFER = ByteBuffer.
                    wrap("hello ServerSocketChannel".getBytes());

    public static void main(String[] args) {
        try (ServerSocketChannel serverSocketChannel =
                                        ServerSocketChannel.open();
             Selector selector = Selector.open()) {
            //设置当前套接字的阻塞方式
            serverSocketChannel.configureBlocking(false);
            //将当前套接字绑定到指定的套接字地址
            serverSocketChannel.bind(new InetSocketAddress(10099));
            System.out.println(serverSocketChannel.validOps());
            //把当前通道注册到指定的选择器中
            SelectionKey selectionKey = serverSocketChannel.
                    register(selector, SelectionKey.OP_ACCEPT, "AAA");
            System.out.println(selectionKey);
            System.out.println(selector.keys());
            while (selector.select() > 0) {
                Set<SelectionKey> selectionKeys = selector.selectedKeys();
                Iterator<SelectionKey> iterator = selectionKeys.iterator();
                while (iterator.hasNext()) {
                    SelectionKey nextKey = iterator.next();
                    System.out.println("消费了:" + nextKey);
                    iterator.remove();
                    System.out.println(nextKey.attachment());
                    if (nextKey.isValid() && nextKey.isAcceptable()) {
                        ServerSocketChannel channel =
                                (ServerSocketChannel) nextKey.channel();
                        //客户端连接完成
                        try (SocketChannel accept = channel.accept()) {
                            if (accept.isConnected() ||
                                            accept.finishConnect()) {
                                //向客户端发送固定数据
                                accept.write(BUFFER);
                                //重置数据指针
                                BUFFER.clear();
                                System.out.println("客户端连接成功:" +
```

```
                        accept.getRemoteAddress());
                    } else {
                        System.out.println("Channel 连接有错误");
                    }
                } catch (Exception e) {
                    e.printStackTrace();
                }
            } else {
                System.out.println("KEY 连接有错误");
            }
        }
    }
    } catch (Exception e) {
        e.printStackTrace();
    }
  }
}
```

执行结果如下：

```
16
channel = sun.nio.ch.ServerSocketChannelImpl[/[0:0:0:0:0:0:0:0]:10099], selector = sun.nio.ch.WEPollSelectorImpl@4361bd48, interestOps = 16, readyOps = 0
[channel = sun.nio.ch.ServerSocketChannelImpl[/[0:0:0:0:0:0:0:0]:10099], selector = sun.nio.ch.WEPollSelectorImpl@4361bd48, interestOps = 16, readyOps = 0]
消费了:channel = sun.nio.ch.ServerSocketChannelImpl[/[0:0:0:0:0:0:0:0]:10099], selector = sun.nio.ch.WEPollSelectorImpl@4361bd48, interestOps = 16, readyOps = 16
AAA
客户端连接成功:/192.168.3.38:58399
消费了:channel = sun.nio.ch.ServerSocketChannelImpl[/[0:0:0:0:0:0:0:0]:10099],
selector = sun.nio.ch.WEPollSelectorImpl@4361bd48, interestOps = 16, readyOps = 16
AAA
客户端连接成功:/192.168.3.38:58400
```

使用 telnet 测试，如图 12-2 所示。

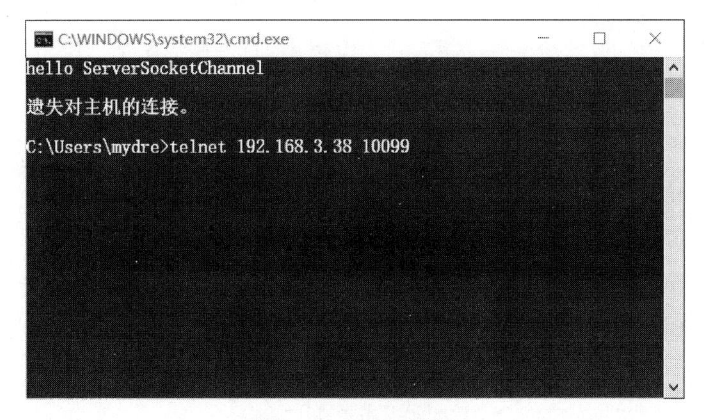

图 12-2 telnet 测试

12.5 Selector 抽象类

Selector 抽象类为可选择信道复用器，配合 ServerSocketChannel、SocketChannel、SelectionKey 使用。

本节基于官方文档介绍常用方法，核心方法配合代码示例以方便读者理解。

1. open()

通过调用系统默认的 SelectorProvider 对象的 openSelector 方法来创建新的选择器对象，代码如下：

```
//第 12 章/five/SelectorTest.java
public class SelectorTest {

    public static void main(String[] args) {
        try (Selector selector = Selector.open()) {
            System.out.println(selector.provider());
        } catch (Exception e) {
            e.printStackTrace();
        }
    }
}
```

执行结果如下：

```
sun.nio.ch.WEPollSelectorProvider@7de26db8
```

此方法的源代码如图 12-3 所示。

```
public static Selector open() throws IOException {
    return SelectorProvider.provider().openSelector();
}
```

图 12-3　源代码

2. close()

关闭当前选择器并释放资源。

3. isOpen()

检查当前选择器是否已打开，返回 boolean 值。

4. keys()

返回当前选择器中的所有 SelectionKey 对象集合。

5. provider()

返回系统默认的 SelectorProvider 对象，代码如下：

```
//第 12 章/five/SelectorTest.java
public class SelectorTest {

    public static void main(String[] args) {
        try (Selector selector = Selector.open()) {
            System.out.println(selector.isOpen());
```

```
            System.out.println(selector.provider());
        } catch (Exception e) {
            e.printStackTrace();
        }
    }
}
```

执行结果如下：

```
true
sun.nio.ch.WEPollSelectorProvider@7de26db8
```

6. select()

选择通道已准备好进行 I/O 操作的 SelectionKey 对象集合，此方法执行分块选择操作，它仅在至少选择一个通道就绪后返回，或者选择器的唤醒方法被调用，或者当前执行线程中断。返回已更新其就绪操作集的数量，代码如下：

```
//第 12 章/five/SelectorTest.java
public class SelectorTest {

    public static void main(String[] args) {
        try (Selector selector = Selector.open()) {
            System.out.println(selector.isOpen());
            System.out.println(selector.provider());
            System.out.println(selector.select());
        } catch (Exception e) {
            e.printStackTrace();
        }
    }
}
```

执行结果如下：

```
true
sun.nio.ch.WEPollSelectorProvider@7de26db8
```

注意：以上代码运行后程序会一直处于阻塞状态。

7. select(long timeout)

选择通道已准备好进行 I/O 操作的 SelectionKey 对象集合，此方法执行分块选择操作，它仅在至少选择一个通道就绪后返回，或者选择器的唤醒方法被调用，或者当前执行线程中断，或者到达了最长超时时间。返回已更新其就绪操作集的数量。接收 long 入参，作为最长超时时间毫秒数，代码如下：

```
//第 12 章/five/SelectorTest.java
public class SelectorTest {
```

```
    public static void main(String[] args) {
        try (Selector selector = Selector.open()) {
            System.out.println(selector.isOpen());
            System.out.println(selector.provider());
            System.out.println(selector.select(5000));
        } catch (Exception e) {
            e.printStackTrace();
        }
    }
}
```

执行结果如下：

```
true
sun.nio.ch.WEPollSelectorProvider@7de26db8
0
```

注意：以上代码运行后程序会在5s后结束。

8. selectNow()

选择通道已准备好进行I/O操作的SelectionKey对象集合。返回已更新其就绪操作集的数量。如果没有任何通道就绪，则此方法立即返回零，代码如下：

```
//第12章/five/SelectorTest.java
public class SelectorTest {

    public static void main(String[] args) {
        try (Selector selector = Selector.open()) {
            System.out.println(selector.isOpen());
            System.out.println(selector.provider());
            System.out.println(selector.selectNow());
        } catch (Exception e) {
            e.printStackTrace();
        }
    }
}
```

执行结果如下：

```
true
sun.nio.ch.WEPollSelectorProvider@7de26db8
0
```

9. selectedKeys()

返回当前选择器中的所有已就绪SelectionKey对象集合。

10. wakeup()

选择器的唤醒方法会唤醒当前的选择器，代码如下：

```java
//第 12 章/five/SelectorTest.java
public class SelectorTest {

    public static void main(String[] args) {
        try (Selector selector = Selector.open()) {
            new Thread(() -> {
                try {
                    Thread.sleep(5000);
                    selector.wakeup();
                } catch (Exception e) {
                    e.printStackTrace();
                }
            }).start();
            System.out.println(selector.isOpen());
            System.out.println(selector.provider());
            System.out.println(selector.select());
        } catch (Exception e) {
            e.printStackTrace();
        }
    }
}
```

执行结果如下：

```
true
sun.nio.ch.WEPollSelectorProvider@1175e2db
0
```

注意：*以上代码运行后程序会在5s后结束。*

12.6 SelectionKey 抽象类

SelectionKey 抽象类表示 SelectableChannel 向选择器注册的令牌。每次向选择器注册一个通道时，就会创建一个选择键，键一直保持有效。直到通过调用其取消方法、关闭其通道、关闭其选择器来取消它为止。

12.6.1 操作集标识

表示通道受支持操作的操作集标识，见表 12-2。

表 12-2 操作集标识

操作集标识	描　述
OP_ACCEPT	套接字接受操作的操作集位
OP_CONNECT	套接字连接操作的操作集位

续表

操作集标识	描　述
OP_READ	套接字读操作的操作集位
OP_WRITE	套接字写操作的操作集位

12.6.2　常用方法

本节基于官方文档介绍常用方法，核心方法配合代码示例以方便读者理解。

1. attach(Object ob)

将指定的对象附加到当前的选择键中。接收 Object 入参，作为指定的对象。

2. attachment()

返回附加到当前选择键中的对象。

3. cancel()

请求取消当前选择键在其选择器中的注册。

4. channel()

返回创建当前选择键的 SelectableChannel 通道。

5. interestOps()

返回当前选择键注册到选择器中的操作集标识，代码如下：

```
//第 12 章/six/SelectionKeyTest.java
public class SelectionKeyTest {

    public static void main(String[] args) {
        try (SocketChannel socketChannel = SocketChannel.open();
             Selector selector = Selector.open()) {
            socketChannel.bind(new InetSocketAddress(10066));
            socketChannel.configureBlocking(false);
            SelectionKey selectionKey = socketChannel.register(selector,
                    SelectionKey.OP_READ | SelectionKey.OP_WRITE);
            System.out.println(selectionKey.interestOps());
            System.out.println(SelectionKey.OP_READ |
                                            SelectionKey.OP_WRITE);

        } catch (Exception e) {
            e.printStackTrace();
        }
    }
}
```

执行结果如下：

```
5
5
```

6. interestOps(int ops)

修改当前选择键注册到选择器中的操作集标识。接收 int 入参，作为操作集标识，代码如下：

```
//第 12 章/six/SelectionKeyTest.java
public class SelectionKeyTest {

    public static void main(String[] args) {
        try (SocketChannel socketChannel = SocketChannel.open();
             Selector selector = Selector.open()) {
            socketChannel.bind(new InetSocketAddress(10066));
            socketChannel.configureBlocking(false);
            SelectionKey selectionKey = socketChannel.register(selector,
                    SelectionKey.OP_READ | SelectionKey.OP_WRITE);
            System.out.println(selectionKey.interestOps());
            System.out.println(SelectionKey.OP_READ |
                                            SelectionKey.OP_WRITE);
            selectionKey.interestOps(SelectionKey.OP_READ);
            System.out.println(selectionKey.interestOps());
        } catch (Exception e) {
            e.printStackTrace();
        }
    }
}
```

执行结果如下：

```
5
5
1
```

7. interestOpsAnd(int ops)

修改当前选择键注册到选择器中的操作集标识。接收 int 入参，作为操作集标识。此方法执行 & 位运算操作。

8. interestOpsOr(int ops)

修改当前选择键注册到选择器中的操作集标识。接收 int 入参，作为操作集标识。此方法执行 | 位运算操作。

9. isAcceptable()

检查当前选择键的通道是否已准备好接受新的套接字连接，返回 boolean 值。

10. isConnectable()

检查当前选择键的通道是否已完成它的套接字连接操作，返回 boolean 值。

11. isReadable()

检查当前选择键的通道是否已准备好进行数据读取，返回 boolean 值。

12．isWritable()

检查当前选择键的通道是否已准备好进行数据写出，返回 boolean 值。

13．isValid()

检查当前选择键是否有效，返回 boolean 值。

14．readyOps()

返回当前选择键注册到选择器中的已就绪操作集标识，代码如下：

```
//第 12 章/six/SelectionKeyTest.java
public class SelectionKeyTest {

    public static void main(String[] args) {
        try (SocketChannel socketChannel = SocketChannel.open();
             Selector selector = Selector.open()) {
            socketChannel.bind(new InetSocketAddress(10066));
            socketChannel.configureBlocking(false);
            SelectionKey selectionKey = socketChannel.register(selector,
                    SelectionKey.OP_READ | SelectionKey.OP_WRITE);
            System.out.println(selectionKey.interestOps());
            System.out.println(SelectionKey.OP_READ |
                                        SelectionKey.OP_WRITE);
            selectionKey.interestOps(SelectionKey.OP_READ);
            System.out.println(selectionKey.interestOps());
            System.out.println(selectionKey.readyOps());
        } catch (Exception e) {
            e.printStackTrace();
        }
    }
}
```

执行结果如下：

```
5
5
1
0
```

15．selector()

返回与当前选择键相关联的选择器对象。

21min

12.7 SocketChannel 抽象类

此类表示客户端套接字通道，通过请求服务器端套接字建立网络连接。当网络连接完成后便可以进行双向通信。

12.7.1　常用方法

本节基于官方文档介绍常用方法，核心方法配合代码示例以方便读者理解。

1. bind(SocketAddress local)

将当前套接字绑定到指定的套接字地址。接收 SocketAddress 入参，作为指定的套接字地址。

2. connect(SocketAddress remote)

将当前套接字连接到指定的套接字地址。接收 SocketAddress 入参，作为指定的套接字地址。

3. finishConnect()

完成当前套接字连接的过程，返回 boolean 值。

4. getLocalAddress()

返回当前套接字所绑定的地址。

5. getRemoteAddress()

返回当前套接字所连接的地址。

6. isConnected()

检查当前套接字是否已经连接，返回 boolean 值。

7. isConnectionPending()

检查当前套接字是否正在连接中，返回 boolean 值。

8. shutdownInput()

关闭当前套接字的输入流。

9. shutdownOutput()

关闭当前套接字的输出流。

10. read(ByteBuffer dst)

从当前套接字读取数据并存储到指定的字节缓冲区，返回读取到的字节总数。如果通道已关闭，则为－1。接收 ByteBuffer 入参，作为指定的字节缓冲区。

11. read(ByteBuffer[] dsts, int offset, int length)

从当前套接字读取数据并存储到指定的字节缓冲区，返回读取到的字节总数。如果通道已关闭，则为－1。接收 ByteBuffer 入参，作为指定的字节缓冲区，接收 int 入参，作为字节缓冲区的起始索引，接收 int 入参，作为最大的数据长度。

12. setOption(SocketOption<T> name, T value)

设置当前套接字指定选项的值。接收 SocketOption 入参，作为指定选项，接收 T 入参，作为指定选项的值。

13. socket()

返回当前的套接字对象。

14. validOps()

返回当前通道受支持操作的操作集标识。

15. configureBlocking(boolean block)

设置当前套接字的阻塞方式。接收 boolean 入参，作为套接字的阻塞方式。

16. write(ByteBuffer src)

将指定的字节缓冲区数据写出到当前套接字，返回写出的字节总数。接收 ByteBuffer 入参，作为指定的字节缓冲区。

17. write(ByteBuffer[] srcs, int offset, int length)

将指定的字节缓冲区数据写出到当前套接字，返回写出的字节总数。接收 ByteBuffer 入参，作为指定的字节缓冲区，接收 int 入参，作为字节缓冲区的起始索引，接收 int 入参，作为最大的数据长度。

18. register(Selector sel, int ops, Object att)

把当前通道注册到指定的选择器中。接收 Selector 入参，作为指定的选择器，接收 int 入参，作为操作集标识，接收 Object 入参，作为附加的数据。返回 SelectionKey 对象。

19. close()

关闭当前套接字通道并释放资源。

20. isOpen()

检查当前通道是否已打开，返回 boolean 值。

21. getOption(SocketOption <T> name)

返回当前套接字指定选项的值。接收 SocketOption 入参，作为指定选项。

22. supportedOptions()

返回当前套接字所支持的套接字选项集合。

23. isBlocking()

检查当前套接字的阻塞方式，返回 boolean 值。

24. isRegistered()

检查当前通道是否向任何选择器注册，返回 boolean 值。

25. keyFor(Selector sel)

返回指定的选择器检索到的当前通道注册的键。接收 Selector 入参，作为指定的选择器。

26. provider()

返回创建当前套接字通道的提供者。

12.7.2 使用示例

使用非阻塞模式设计一个客户端,连接到指定的服务器端,连接完成后接收服务器端发送的数据,代码如下:

```
//第 12 章/seven/SocketChannelTest.java
//import 第 12 章/four/ServerSocketChannelTest.java
public class SocketChannelTest {

    private static final ByteBuffer BUFFER = ByteBuffer.allocate(1024);

    public static void main(String[] args) {
        try (SocketChannel socketChannel = SocketChannel.open();
             Selector selector = Selector.open()) {
            socketChannel.configureBlocking(false);
            socketChannel.bind(new InetSocketAddress(10066));
            socketChannel.connect(new InetSocketAddress(10099));
            System.out.println(socketChannel.validOps());
            if (socketChannel.isConnected() ||
                            socketChannel.finishConnect()) {
                socketChannel.register(selector, SelectionKey.OP_READ);
                while (selector.select() > 0) {
                    Set<SelectionKey> selectionKeys = selector.selectedKeys();
                    Iterator<SelectionKey> iterator =
                                            selectionKeys.iterator();
                    while (iterator.hasNext()) {
                        SelectionKey next = iterator.next();
                        iterator.remove();
                        if (next.isValid() && next.isReadable()) {
                            SocketChannel readChannel =
                                        (SocketChannel) next.channel();
                            int read = readChannel.read(BUFFER);
                            BUFFER.flip();
                            byte[] bts = new byte[BUFFER.remaining()];
                            BUFFER.get(bts);
                            System.out.println(new String(bts));
                            BUFFER.compact();
                            //套接字通道关闭
                            if (read == -1) {
                                next.cancel();
                                return;
                            }
                        }
                    }
                }
            }
```

```
            } else {
                System.out.println("连接失败...");
            }
        } catch (Exception e) {
            e.printStackTrace();
        }
    }
}
```

执行结果如下：

```
13
hello ServerSocketChannel
```

小结

本章介绍了 NIO 包下网络套接字的使用，读者应理解 NIO 和 BIO 使用方式的区别。

习题

1. 判断题

(1) ServerSocketChannel 表示服务器端套接字通道是一个抽象类。(　　)

(2) SocketChannel 表示客户端套接字通道是一个抽象类。(　　)

(3) ServerSocketChannel 只能注册 SelectionKey. OP_ACCEPT 操作集标识。(　　)

2. 选择题

(1) 表示套接字通道的类有(　　)。(多选)

A. ServerSocketChannel 类　　B. SocketChannel 类

C. Socket 类　　D. ServerSocket 类

(2) 设置套接字通道阻塞方式的方法是(　　)。(单选)

A. configureBlocking(boolean block)　　B. isBlocking()

C. validOps()　　D. register(Selector sel, int ops)

(3) 表示套接字通道受支持操作的操作集标识方法是(　　)。(单选)

A. configureBlocking(boolean block)　　B. isBlocking()

C. validOps()　　D. register(Selector sel, int ops)

(4) 表示套接字通道注册到指定的选择器方法是(　　)。(单选)

A. configureBlocking(boolean block)　　B. isBlocking()

C. validOps()　　D. register(Selector sel, int ops)

(5) 表示设置套接字通道指定选项的值方法是(　　)。(单选)

A. configureBlocking(boolean block)

B. setOption(SocketOption<T> name, T value)

C. validOps()

D. register(Selector sel, int ops)

3. 填空题

(1) 查看执行结果并补充代码,代码如下:

```
//第 12 章/answer/OneAnswer.java
public class OneAnswer {

    public static void main(String[] args) {
        try (Selector selector = Selector.open()) {
            new Thread(() ->{
                try {
                    Thread.sleep(5000);
                    selector.________;
                } catch (Exception e) {
                    e.printStackTrace();
                }
            }).start();
            System.out.println(selector.isOpen());
            System.out.println(selector.provider());
            System.out.println(selector.select());
        } catch (Exception e) {
            e.printStackTrace();
        }
    }
}
```

执行结果如下:

```
true
sun.nio.ch.WEPollSelectorProvider@36aa7bc2
0
```

(2) 根据业务要求补全代码,使用非阻塞模式设计一个服务器端。接收客户端的连接,连接完成后向客户端发送固定数据,数据发送完毕后关闭此次的客户端连接,代码如下:

```
//第 12 章/answer/TwoAnswer.java
public class TwoAnswer {

    private static final ByteBuffer BUFFER = ByteBuffer.wrap
                                    (_________________________.getBytes());

    public static void main(String[] args) {
        try (ServerSocketChannel serverSocketChannel =
                                        ServerSocketChannel.open();
            Selector selector = Selector.open()) {
```

```
            //设置当前套接字的阻塞方式
            serverSocketChannel.configureBlocking(________);
            //将当前套接字绑定到指定的套接字地址
            serverSocketChannel.bind(new InetSocketAddress(10099));
            //把当前通道注册到指定的选择器中
            serverSocketChannel.register(selector, ________);
            while (selector.select() > 0) {
                Set<SelectionKey> selectionKeys = selector.________;
                Iterator<SelectionKey> iterator = selectionKeys.iterator();
                while (iterator.hasNext()) {
                    SelectionKey nextKey = iterator.next();
                    System.out.println("消费了:" + nextKey);
                    iterator.remove();
                    System.out.println(nextKey.attachment());
                    if (nextKey.isValid() && nextKey.isAcceptable()) {
                        ServerSocketChannel channel =
                                (ServerSocketChannel) nextKey.channel();
                        /客户端连接完成
                        try (SocketChannel accept = channel.accept()) {
                            if (accept.isConnected() ||
                                        accept.finishConnect()) {
                                //向客户端发送固定数据
                                accept.write(BUFFER);
                                //重置数据指针
                                BUFFER.clear();
                                System.out.println("客户端连接成功:" +
                                        accept.getRemoteAddress());
                            } else {
                                System.out.println("Channel 连接有错误");
                            }
                        } catch (Exception e) {
                            e.printStackTrace();
                        }
                    } else {
                        System.out.println("KEY 连接有错误");
                    }
                }
            }
        } catch (Exception e) {
            e.printStackTrace();
        }
    }
}
```

执行结果如下：

```
消费了:channel = sun.nio.ch.ServerSocketChannelImpl[/[0:0:0:0:0:0:0:0]:10099], selector =
sun.nio.ch.WEPollSelectorImpl@762efe5d, interestOps = 16, readyOps = 16
null
客户端连接成功:/192.168.3.38:63212
```

第13章 泛　型

CHAPTER 13

泛型由JDK 1.5增加，通过泛型可以以类型安全的方式使用各种类型数据，可以增强源代码的可阅读性，减少可能产生的由类型强转造成的错误。

13.1 泛型类声明

声明任意一个类中包装一个任意类型的数据，并提供构造器入参和标准的get、set方法。

13.1.1 普通类演示

基于普通类的演示，代码如下：

```
//第13章/one/GenericTest.java
public class GenericTest {

    public static void main(String[] args) {
        GenericBase g1 = new GenericBase("狗");
        String str = (String) g1.getObject();
        System.out.println(str);

        GenericBase g2 = new GenericBase(18);
        Integer i1 = (Integer) g2.getObject();
        System.out.println(i1);
    }

    public static final class GenericBase{

        private Object object;

        public GenericBase(Object object) {
            this.object = object;
        }
```

```
        public Object getObject() {
            return object;
        }

        public void setObject(Object object) {
            this.object = object;
        }
    }
}
```

执行结果如下：

```
狗
18
```

13.1.2 泛型类演示

基于泛型类的演示，代码如下：

```
//第 13 章/one/GenericTest.java
public class GenericTest {

    public static void main(String[] args) {
        GenericTwo<String> g3 = new GenericTwo<>("狗");
        String e = g3.getE();
        System.out.println(e);

        GenericTwo<Integer> g4 = new GenericTwo<>(18);
        Integer i2 = g4.getE();
        System.out.println(i2);
    }

    public static final class GenericTwo<E>{
        private E e;

        public GenericTwo(E e) {
            this.e = e;
        }

        public E getE() {
            return e;
        }

        public void setE(E e) {
            this.e = e;
        }
    }

}
```

执行结果如下：

```
狗
18
```

13.2 泛型类型限制

泛型类型在使用时，可以限制泛型类型的范围。

13.2.1 固定泛型类型

固定泛型类型限制，代码如下：

```java
//第 13 章/two/GenericControl.java
public class GenericControl {

    public static void main(String[] args) {
        GenericTwo<String> genericTwo = new GenericTwo<>("泛型类型");
        callOne(genericTwo);

        GenericTwo<Integer> genericInt = new GenericTwo<>(18);
        callTwo(genericInt);

    }
    //固定泛型类型
    public static void callOne(GenericTwo<String> genericTwo){
        String e = genericTwo.getE();
        System.out.println(e);
    }
    //固定泛型类型
    public static void callTwo(GenericTwo<Integer> genericTwo){
        Integer e = genericTwo.getE();
        System.out.println(e);
    }


    public static final class GenericTwo<E>{
        private E e;

        public GenericTwo(E e) {
            this.e = e;
        }

        public E getE() {
            return e;
        }
```

```
        public void setE(E e) {
            this.e = e;
        }
    }

}
```

执行结果如下：

```
泛型类型
18
```

13.2.2 通用泛型类型

通用泛型类型相当于 Object 类型，代码如下：

```
//第 13 章/two/GenericControl.java
public class GenericControl {

    public static void main(String[] args) {
        GenericTwo<String> genericTwo = new GenericTwo<>("泛型类型");
        GenericTwo<Integer> genericInt = new GenericTwo<>(18);
        callAll(genericTwo);
        callAll(genericInt);
    }

    //通用泛型类型
    public static void callAll(GenericTwo<?> genericTwo){
        if(genericTwo.getE() instanceof String){
            String str = (String) genericTwo.getE();
            System.out.println(str);
        }
        Object e = genericTwo.getE();
        System.out.println(e);
    }

    public static final class GenericTwo<E>{
        private E e;

        public GenericTwo(E e) {
            this.e = e;
        }

        public E getE() {
            return e;
        }

        public void setE(E e) {
```

```
            this.e = e;
        }
    }

}
```

执行结果如下：

```
泛型类型
泛型类型
18
```

13.2.3 泛型上限控制

泛型上限控制可以控制泛型类型的最大范围，代码如下：

```
//第 13 章/two/GenericControl.java
public class GenericControl {

    public static void main(String[] args) {
        GenericTwo<Integer> genericInt = new GenericTwo<>(18);
        callThree(genericInt);

        GenericTwo<Float> genericFlt = new GenericTwo<>(18.66f);
        callThree(genericFlt);
    }
    //泛型上限控制
    public static void callThree(GenericTwo<? extends Number> genericTwo){
        Number e = genericTwo.getE();
        System.out.println(e);
    }

    public static final class GenericTwo<E>{
        private E e;

        public GenericTwo(E e) {
            this.e = e;
        }

        public E getE() {
            return e;
        }

        public void setE(E e) {
            this.e = e;
        }
    }

}
```

执行结果如下：

```
18
18.66
```

13.2.4 泛型下限控制

泛型下限控制可以控制泛型类型的最小范围，代码如下：

```
//第13章/two/GenericControl.java
public class GenericControl {

    public static void main(String[] args) {
        GenericTwo<Integer> genericInt = new GenericTwo<>(18);
        callFour(genericInt);
        //Float 转型成 Number
        GenericTwo<Number> genericFlt = new GenericTwo<>(18.66f);
        callFour(genericFlt);
    }
    //泛型下限控制
    public static void callFour(GenericTwo<? super Integer> genericTwo){
        Object e = genericTwo.getE();
        System.out.println(e);
    }

    public static final class GenericTwo<E>{
        private E e;

        public GenericTwo(E e) {
            this.e = e;
        }

        public E getE() {
            return e;
        }

        public void setE(E e) {
            this.e = e;
        }
    }

}
```

执行结果如下：

```
18
18.66
```

13.3 泛型声明的几种方式

13.3.1 泛型类

泛型类声明方式，用于对象级别的管控，代码如下：

```
//第 13 章/three/GenericStatement.java
public class GenericStatement {
    public static void main(String[] args) {
        GenericTwo<Integer> genericInt = new GenericTwo<>(18);
        GenericTwo<String> genericStr = new GenericTwo<>("str");
        System.out.println(genericInt.getE());
        System.out.println(genericStr.getE());
    }


    public static final class GenericTwo<E>{
        private E e;

//private static E ste; 错误

        public GenericTwo(E e) {
            this.e = e;
        }

        public E getE() {
            return e;
        }

        public void setE(E e) {
            this.e = e;
        }
//public static void callOne(E e) {} 错误

    }
}
```

执行结果如下：

```
18
str
```

13.3.2 泛型静态方法

可以直接在静态方法上声明使用泛型，一般用来控制入参和返回类型，代码如下：

```java
//第 13 章/three/GenericStatement.java
public class GenericStatement {

    public static void main(String[] args) {
        String str = callOne("泛型声明静态方法");
        System.out.println(str);
        Integer integer = callOne(18);
        System.out.println(integer);
    }

    public static <E> E callOne(E obj){
        return obj;
    }

}
```

执行结果如下：

```
泛型声明静态方法
18
```

13.3.3 泛型对象方法

可以直接在对象方法上声明使用泛型，一般用来控制入参和返回类型，代码如下：

```java
//第 13 章/three/GenericStatement.java
public class GenericStatement {

    public static void main(String[] args) {
        GenericStatement genericStatement = new GenericStatement();
        String str = genericStatement.callThree("泛型对象方法");
        Integer num = genericStatement.callThree(18);
        System.out.println(str);
        System.out.println(num);
    }

    public <E> E callThree(E obj){
        return obj;
    }

}
```

执行结果如下：

```
泛型对象方法
18
```

小结

通过泛型可以类型安全的方式使用各种类型数据，可以增强源代码的可阅读性，减少可能产生的由类型强转造成的错误。

习题

1. 判断题

(1) 泛型可以减少可能产生的由类型强转造成的错误。(　　)

(2) 泛型上限控制可以控制泛型类型的最大范围。(　　)

(3) 泛型下限控制可以控制泛型类型的最小范围。(　　)

2. 选择题

(1) 泛型上限控制的语法是(　　)。(单选)

A. List <? extends Number >　　B. List < Number >

C. List <? >　　D. List < Object >

(2) 泛型下限控制的语法是(　　)。(单选)

A. List <? super Integer >　　B. List < Number >

C. List <? >　　D. List < Object >

(3) 泛型不限制类型的语法有(　　)。(多选)

A. List <? super Integer >　　B. List < Number >

C. List <? >　　D. List < Object >

3. 填空题

查看执行结果并补充代码，代码如下：

```
//第 13 章/answer/ElevenOne.java
public class ElevenOne {
    public static void main(String[ ] args) {
        Generic < String > genericOne = new Generic <>("泛型类型");
        callOne(genericOne);
        Generic < Integer > genericInt = new Generic <>(18);
        callTwo(genericInt);

        callAll(genericOne);
        callAll(genericInt);
    }

    public static void callOne(Generic <________> generic){
        String e = generic.getE();
```

```
            System.out.println(e);
        }

        public static void callTwo(Generic<? ________ Number> generic){
            Number e = generic.getE();
            System.out.println(e);
        }

        public static void callAll(Generic<________> generic){
            if(generic.getE() instanceof String){
                String e = (String) generic.getE();
                System.out.println(e);
            }
            Object e = generic.getE();
            System.out.println(e);
        }

        public static final class Generic<E>{
            private E e;

            public Generic(E e) {
                this.e = e;
            }

            public E getE() {
                return e;
            }

            public void setE(E e) {
                this.e = e;
            }
        }
}
```

执行结果如下：

```
泛型类型
18
泛型类型
泛型类型
18
```

第 14 章

CHAPTER 14

List 集合框架

List 接口继承了 Collection 接口，List 接口被称为有序集合框架。此接口定义了数据的基础操作功能，底层有多个实现类，例如 ArrayList、LinkedList、Vector、Stack 等。

14.1 List 接口

有序集合框架此接口的每个元素的插入位置具有精确控制权。本节基于官方文档介绍常用方法，核心方法配合代码示例以方便读者理解。

1. add(int index, E element)

将指定的数据插入此集合中指定的位置。接收 int 入参，作为指定位置，接收 E 入参，作为指定数据。

2. add(E e)

将指定的数据添加到此集合的尾部，返回 boolean 值。接收 E 入参，作为指定数据。

3. addAll(int index, Collection <? extends E > c)

将指定的集合数据插入此集合指定的位置，返回 boolean 值。接收 int 入参，作为指定位置，接收 Collection 入参，作为指定集合数据。

4. addAll(Collection <? extends E > c)

将指定的集合数据添加到此集合的尾部，返回 boolean 值。接收 Collection 入参，作为指定集合数据。

5. clear()

清空此集合内的所有数据。

6. contains(Object o)

检查此集合中是否包含指定的数据，返回 boolean 值。接收 Object 入参，作为指定数据。

7. containsAll(Collection<?> c)

检查此集合中是否包含指定的集合数据，返回 boolean 值。接收 Collection 入参，作为指定集合数据。

8. equals(Object o)

将此对象与指定对象进行比较，返回 boolean 值。接收 Object 入参，作为指定对象。

9. get(int index)

返回此集合中指定位置的数据。接收 int 入参，作为指定位置。

10. indexOf(Object o)

返回此集合中首次出现的指定数据的索引，如果此集合不包含该数据，则返回－1。

11. isEmpty()

检查此集合是否为空，返回 boolean 值。

12. lastIndexOf(Object o)

返回此集合中最后一次出现的指定数据的索引，如果此集合不包含该数据，则返回－1。

13. remove(int index)

删除此集合中指定位置的数据，并返回删除的数据。接收 int 入参，作为指定位置。

14. remove(Object o)

从此集合中删除首次出现的指定数据，返回 boolean 值。接收 Object 入参，作为指定数据。

15. removeAll(Collection<?> c)

从此集合中删除指定集合中包含的所有数据，返回 boolean 值。接收 Collection 入参，作为指定集合数据。

16. replaceAll(UnaryOperator<E> operator)

根据指定的数据规则从此集合中替换所有数据。接收 UnaryOperator 入参，作为指定数据规则。

17. set(int index, E element)

将此集合中指定位置的数据替换为指定的数据，并返回旧数据。接收 int 入参，作为指定位置，接收 E 入参，作为指定数据。

18. size()

返回此集合中数据的长度。

19. sort(Comparator<? super E> c)

根据指定的比较器所引导的顺序对该集合数据进行排序。接收 Comparator 入参，作为指定的比较器。

20. subList(int fromIndex, int toIndex)

从指定的开始位置到指定的结束位置分割此集合数据，并返回新的集合对象。接收 int 入参，作为开始位置，接收 int 入参，作为结束位置。

21. removeIf(Predicate <? super E > filter)

根据指定的条件规则从此集合中删除数据。接收 Predicate 入参，作为指定条件规则。

22. forEach(Consumer <? super T > action)

根据指定的消费规则从此集合中遍历所有数据。接收 Consumer 入参，作为指定消费规则。

23. of(E... elements)

静态方法，从指定的数据中创建此集合对象。接收 E 泛型入参，作为指定数据，代码如下：

```java
//第 14 章/one/ListTest.java
public class ListTest {

    public static void main(String[] args) {
        //ListN<E>实现类,不允许修改数据
        List<Integer> integers = List.of(15, 16, 17, 18,
                                        19, 20,15, 16, 17, 18, 19, 20);
        System.out.println(integers);
//System.out.println(integers.add(888)); error
//integers.set(5,886); error

        System.out.println(integers.contains(15));

        System.out.println(integers.size());
        System.out.println(integers.get(11));

        //SubList<E>实现类
        List<Integer> subList = integers.subList(3, 10);
        System.out.println(subList);

    }
}
```

执行结果如下：

```
[15, 16, 17, 18, 19, 20, 15, 16, 17, 18, 19, 20]
true
12
20
[18, 19, 20, 15, 16, 17, 18]
```

此种方法创建的默认实现类都继承了 AbstractImmutableList < E >类，部分方法使用受限，如图 14-1 所示。

```
@jdk.internal.ValueBased
static abstract class AbstractImmutableList<E> extends AbstractImmutableCollection<E>
implements List<E>, RandomAccess {

    // all mutating methods throw UnsupportedOperationException
    @Override public void    add(int index, E element) { throw uoe(); }
    @Override public boolean addAll(int index, Collection<? extends E> c) { throw uoe(); }
    @Override public E       remove(int index) { throw uoe(); }
    @Override public void    replaceAll(UnaryOperator<E> operator) { throw uoe(); }
    @Override public E       set(int index, E element) { throw uoe(); }
    @Override public void    sort(Comparator<? super E> c) { throw uoe(); }
    //省略...

}
```

图 14-1 AbstractImmutableList < E >类的源代码

42min

14.2 ArrayList 类

此类实现了 List 接口,底层基于数组实现,具有自动扩容、允许重复数据、允许空数据的特性。

14.2.1 构造器

ArrayList 类构造器,见表 14-1。

表 14-1 ArrayList 类构造器

构 造 器	描 述
ArrayList()	构造新的对象,默认为无参构造器
ArrayList(int initialCapacity)	构造新的对象,指定初始容量
ArrayList(Collection <? extends E > c)	构造新的对象,指定初始数据

14.2.2 常用方法

本节基于官方文档介绍常用方法,核心方法配合代码示例以方便读者理解。

1. add(int index, E element)

将指定的数据插入此集合中指定的位置。接收 int 入参,作为指定位置,接收 E 入参,作为指定数据,代码如下:

```
//第 14 章/two/ArrayListTest.java
public class ArrayListTest {

    public static void main(String[] args) {
        ArrayList < Integer > arrayList = new ArrayList <>(16);
        arrayList.add(15);
```

```
        arrayList.add(16);
        arrayList.add(17);
        arrayList.add(18);
        System.out.println(arrayList);              //[15, 16, 17, 18]
        arrayList.add(2,886);
        System.out.println(arrayList);              //[15, 16, 886, 17, 18]
    }
}
```

执行结果如下：

```
[15, 16, 17, 18]
[15, 16, 886, 17, 18]
```

2. add(E e)

将指定的数据添加到此集合的尾部,返回 boolean 值。接收 E 入参,作为指定数据。

3. addAll(int index, Collection <? extends E > c)

将指定的集合数据插入此集合指定的位置,返回 boolean 值。接收 int 入参,作为指定位置,接收 Collection 入参,作为指定集合数据。

4. addAll(Collection <? extends E > c)

将指定的集合数据添加到此集合的尾部,返回 boolean 值。接收 Collection 入参,作为指定集合数据。

5. clear()

清空此集合内的所有数据,代码如下：

```
//第 14 章/two/ArrayListTest.java
public class ArrayListTest {

    public static void main(String[] args) {
        ArrayList < Integer > arrayList  =  new ArrayList <>(16);
        arrayList.add(15);
        arrayList.add(16);
        arrayList.add(17);
        arrayList.add(18);
        System.out.println(arrayList);              //[15, 16, 17, 18]
        arrayList.clear();
        System.out.println(arrayList);              //[]
    }
}
```

执行结果如下：

```
[15, 16, 17, 18]
[]
```

6. contains(Object o)

检查此集合中是否包含指定的数据，返回 boolean 值。接收 Object 入参，作为指定数据。

7. containsAll(Collection<?> c)

检查此集合中是否包含指定的集合数据，返回 boolean 值。接收 Collection 入参，作为指定集合数据，代码如下：

```
//第 14 章/two/ArrayListTest.java
public class ArrayListTest {

    public static void main(String[] args) {
        ArrayList<Integer> arrayList2 = new ArrayList<>(16);
        arrayList2.add(15);
        arrayList2.add(16);
        arrayList2.add(15);
        arrayList2.add(18);
        ArrayList<Integer> arrayList3 = new ArrayList<>(16);
        arrayList3.add(15);
        arrayList3.add(16);
        arrayList3.add(15);
        arrayList3.add(18);
        arrayList3.add(18);
        System.out.println(arrayList2.containsAll(arrayList3));
    }
}
```

执行结果如下：

```
true
```

8. equals(Object o)

对此对象与指定对象进行比较，返回 boolean 值。接收 Object 入参，作为指定对象，代码如下：

```
//第 14 章/two/ArrayListTest.java
public class ArrayListTest {

    public static void main(String[] args) {
        ArrayList<Integer> arrayList2 = new ArrayList<>(16);
        arrayList2.add(15);
        arrayList2.add(16);
        arrayList2.add(15);
        arrayList2.add(18);
        ArrayList<Integer> arrayList3 = new ArrayList<>(16);
        arrayList3.add(15);
        arrayList3.add(16);
        arrayList3.add(15);
```

```
        arrayList3.add(18);
        arrayList3.add(18);
        System.out.println(arrayList2.equals(arrayList3));
    }
}
```

执行结果如下：

```
false
```

9. get(int index)

返回此集合中指定位置的数据。接收 int 入参，作为指定位置。

10. indexOf(Object o)

返回此集合中首次出现的指定数据的索引，如果此集合不包含该数据，则返回－1，代码如下：

```
//第 14 章/two/ArrayListTest.java
public class ArrayListTest {

    public static void main(String[] args) {
        ArrayList<Integer> arrayList = new ArrayList<>(16);
        arrayList.add(15);
        arrayList.add(886);
        arrayList.add(17);
        arrayList.add(18);
        System.out.println(arrayList.indexOf(886));
        System.out.println(arrayList.indexOf(1212121));
    }
}
```

执行结果如下：

```
1
-1
```

11. isEmpty()

检查此集合是否为空，返回 boolean 值。

12. lastIndexOf(Object o)

返回此集合中最后一次出现的指定数据的索引，如果此集合不包含该数据，则返回－1，代码如下：

```
//第 14 章/two/ArrayListTest.java
public class ArrayListTest {

    public static void main(String[] args) {
        ArrayList<Integer> arrayList = new ArrayList<>(16);
```

```
        arrayList.add(15);
        arrayList.add(886);
        arrayList.add(17);
        arrayList.add(18);
        arrayList.add(886);
        arrayList.add(6666);
        System.out.println(arrayList.indexOf(886));
        System.out.println(arrayList.lastIndexOf(886));
    }
}
```

执行结果如下：

```
1
4
```

13. remove(int index)

删除此集合中指定位置的数据，并返回删除的数据。接收 int 入参，作为指定位置，代码如下：

```
//第 14 章/two/ArrayListTest.java
public class ArrayListTest {

    public static void main(String[] args) {
        ArrayList<Integer> arrayList = new ArrayList<>(16);
        arrayList.add(15);
        arrayList.add(886);
        arrayList.add(17);
        arrayList.add(18);
        arrayList.add(886);
        arrayList.add(6666);
        System.out.println(arrayList);
        System.out.println(arrayList.remove(1));
        System.out.println(arrayList);
    }
}
```

执行结果如下：

```
[15, 886, 17, 18, 886, 6666]
886
[15, 17, 18, 886, 6666]
```

14. remove(Object o)

从此集合中删除首次出现的指定数据，返回 boolean 值。接收 Object 入参，作为指定数据。

15. removeAll(Collection<?> c)

从此集合中删除指定集合中包含的所有数据，返回 boolean 值。接收 Collection 入参，作为指定集合数据。

16. replaceAll(UnaryOperator < E > operator)

根据指定的数据规则从此集合中替换所有数据。接收 UnaryOperator 入参,作为指定数据规则,代码如下:

```java
//第 14 章/two/ArrayListTest.java
public class ArrayListTest {

    public static void main(String[] args) {
        ArrayList < Integer > arrayList = new ArrayList <>(16);
        arrayList.add(15);
        arrayList.add(16);
        arrayList.add(17);
        arrayList.add(18);
        System.out.println(arrayList);          //[15, 16, 17, 18]
        arrayList.replaceAll(new UnaryOperator < Integer >() {
            @Override
            public Integer apply(Integer integer) {
                return integer * 2;
            }
        });
        System.out.println(arrayList);          //[30, 32, 34, 36]
    }
}
```

执行结果如下:

```
[15, 16, 17, 18]
[30, 32, 34, 36]
```

17. set(int index, E element)

将此集合中指定位置的数据替换为指定的数据,并返回原数据。接收 int 入参,作为指定位置,接收 E 入参,作为指定数据,代码如下:

```java
//第 14 章/two/ArrayListTest.java
public class ArrayListTest {

    public static void main(String[] args) {
        ArrayList < Integer > arrayList = new ArrayList <>(16);
        arrayList.add(15);
        arrayList.add(16);
        arrayList.add(17);
        arrayList.add(18);
        System.out.println(arrayList);
        System.out.println(arrayList.set(1, 888888));
        System.out.println(arrayList);
    }
}
```

执行结果如下：

```
[15, 16, 17, 18]
16
[15, 888888, 17, 18]
```

18. size()

返回此集合中数据的长度。

19. sort(Comparator <? super E > c)

根据指定的比较器所引导的顺序对该集合数据进行排序。接收 Comparator 入参，作为指定的比较器，代码如下：

```
//第 14 章/two/ArrayListTest.java
public class ArrayListTest {

    public static void main(String[] args) {
        ArrayList < Integer > arrayList = new ArrayList <>(16);
        arrayList.add(152);
        arrayList.add(16);
        arrayList.add(1722);
        arrayList.add(1811);
        arrayList.add(3);
        System.out.println(arrayList);          //[152, 16, 1722, 1811, 3]
        arrayList.sort(new Comparator < Integer >() {
            @Override
            public int compare(Integer o1, Integer o2) {
                return o1 - o2;
            }
        });
        System.out.println(arrayList);          //[3, 16, 152, 1722, 1811]
    }
}
```

执行结果如下：

```
[152, 16, 1722, 1811, 3]
[3, 16, 152, 1722, 1811]
```

20. subList(int fromIndex, int toIndex)

从指定的开始位置到指定的结束位置分割此集合数据，并返回新的集合对象。接收 int 入参，作为开始位置，接收 int 入参，作为结束位置。

21. removeIf(Predicate <? super E > filter)

根据指定的条件规则从此集合中删除数据。接收 Predicate 入参，作为指定条件规则，代码如下：

```java
//第 14 章/two/ArrayListTest.java
public class ArrayListTest {

    public static void main(String[] args) {
        ArrayList<Integer> arrayList = new ArrayList<>(16);
        arrayList.add(152);
        arrayList.add(16);
        arrayList.add(1722);
        arrayList.add(1811);
        arrayList.add(3);
        System.out.println(arrayList);        //[152, 16, 1722, 1811, 3]
        arrayList.removeIf(new Predicate<Integer>() {
            @Override
            public boolean test(Integer integer) {
                return integer.equals(1722);
            }
        });
        System.out.println(arrayList);        //[152, 16, 1811, 3]
    }
}
```

执行结果如下：

```
[152, 16, 1722, 1811, 3]
[152, 16, 1811, 3]
```

22. forEach(Consumer<? super T> action)

根据指定的消费规则从此集合中遍历所有数据。接收 Consumer 入参，作为指定消费规则，代码如下：

```java
//第 14 章/two/ArrayListTest.java
public class ArrayListTest {

    public static void main(String[] args) {
        ArrayList<Integer> arrayList = new ArrayList<>(16);
        arrayList.add(152);
        arrayList.add(16);
        arrayList.add(1722);
        arrayList.add(1811);
        arrayList.add(3);
        //JDK 8 方法引用
        arrayList.forEach(System.out::println);
    }
}
```

执行结果如下：

```
152
16
```

```
1722
1811
3
```

23. trimToSize()

将此集合容器的容量缩减为此集合数据的长度。

31min

14.3 LinkedList 类

此类实现了 List 接口和 Deque 接口，底层基于链表实现，具有允许重复数据、允许空数据的特性，不存在扩容的概念。

数据结构概念，如图 14-2 所示。

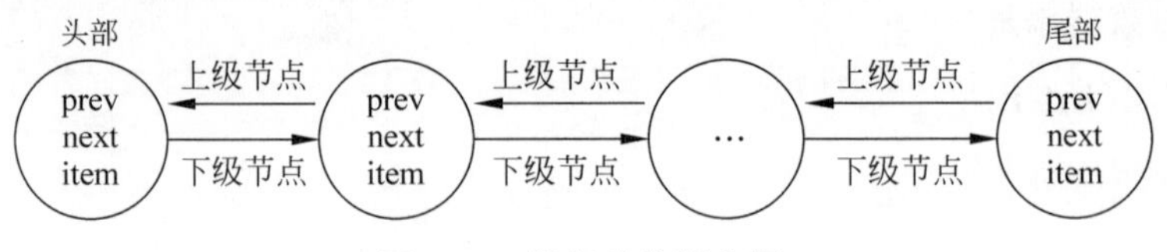

图 14-2 数据结构概念图

14.3.1 构造器

LinkedList 类构造器，见表 14-2。

表 14-2 LinkedList 类构造器

构 造 器	描 述
LinkedList()	构造新的对象，默认为无参构造器
LinkedList(Collection <? extends E > c)	构造新的对象，指定初始数据

14.3.2 常用方法

本节基于官方文档介绍常用方法，核心方法配合代码示例以方便读者理解。

1. add(int index, E element)

将指定的数据插入此集合中指定的位置。接收 int 入参，作为指定位置，接收 E 入参，作为指定数据，代码如下：

```java
//第 14 章/three/LinkedListTest.java
public class LinkedListTest {

    public static void main(String[] args) {
        LinkedList < Integer > linkedList = new LinkedList <>();
        linkedList.add(18);
```

```
        linkedList.add(1545);
        linkedList.add(455);
        linkedList.add(17848);
        linkedList.add(655);
        linkedList.add(2,8888);
        System.out.println(linkedList);
    }

}
```

执行结果如下：

```
[18, 1545, 8888, 455, 17848, 655]
```

2. add(E e)

将指定的数据添加到此集合的尾部，返回 boolean 值。接收 E 入参，作为指定数据。

3. addAll(int index, Collection <? extends E > c)

将指定的集合数据插入此集合指定的位置，返回 boolean 值。接收 int 入参，作为指定位置，接收 Collection 入参，作为指定集合数据。

4. addAll(Collection <? extends E > c)

将指定的集合数据添加到此集合的尾部，返回 boolean 值。接收 Collection 入参，作为指定集合数据。

5. clear()

清空此集合内的所有数据。

6. contains(Object o)

检查此集合中是否包含指定的数据，返回 boolean 值。接收 Object 入参，作为指定数据。

7. containsAll(Collection <?> c)

检查此集合中是否包含指定的集合数据，返回 boolean 值。接收 Collection 入参，作为指定集合数据。

8. equals(Object o)

对此对象与指定对象进行比较，返回 boolean 值。接收 Object 入参，作为指定对象。

9. get(int index)

返回此集合中指定位置的数据。接收 int 入参，作为指定位置。

10. indexOf(Object o)

返回此集合中首次出现的指定数据的索引，如果此集合不包含该数据，则返回－1。

11. isEmpty()

检查此集合是否为空，返回 boolean 值。

12. lastIndexOf(Object o)

返回此集合中最后一次出现的指定数据的索引，如果此集合不包含该数据，则返回－1。

13. remove(int index)

删除此集合中指定位置的数据，并返回删除的数据。接收 int 入参，作为指定位置。

14. remove(Object o)

从此集合中删除首次出现的指定数据，返回 boolean 值。接收 Object 入参，作为指定数据。

15. removeAll(Collection <? > c)

从此集合中删除指定集合中包含的所有数据，返回 boolean 值。接收 Collection 入参，作为指定集合数据。

16. replaceAll(UnaryOperator < E > operator)

根据指定的数据规则从此集合中替换所有数据。接收 UnaryOperator 入参，作为指定数据规则。

17. set(int index, E element)

将此集合中指定位置的数据替换为指定的数据，并返回原数据。接收 int 入参，作为指定位置，接收 E 入参，作为指定数据。

18. size()

返回此集合中数据的长度。

19. sort(Comparator <? super E > c)

根据指定的比较器所引导的顺序对该集合数据进行排序。接收 Comparator 入参，作为指定的比较器。

20. subList(int fromIndex, int toIndex)

从指定的开始位置到指定的结束位置分割此集合数据，并返回新的集合对象。接收 int 入参，作为开始位置，接收 int 入参，作为结束位置。

21. removeIf(Predicate <? super E > filter)

根据指定的条件规则从此集合中删除数据。接收 Predicate 入参，作为指定条件规则。

22. trimToSize()

将此集合容器的容量缩减为此集合数据的长度。

23. addFirst(E e)

将指定的数据添加到此集合的头部。接收 E 入参，作为指定数据，代码如下：

```java
//第 14 章/three/LinkedListTest.java
public class LinkedListTest {

    public static void main(String[] args) {
        LinkedList<Integer> linkedList = new LinkedList<>();
        linkedList.add(18);
        linkedList.add(1545);
        linkedList.add(455);
        linkedList.add(17848);
        linkedList.add(655);
        linkedList.addFirst(8888);                //Deque 接口的方法
        System.out.println(linkedList);
    }

}
```

执行结果如下：

```
[8888, 18, 1545, 455, 17848, 655]
```

24. addLast(E e)

将指定的数据添加到此集合的尾部。接收 E 入参，作为指定数据。

25. getFirst()

返回此集合头部位置的数据，如果数据为空，则抛出异常。

26. getLast()

返回此集合尾部位置的数据，如果数据为空，则抛出异常。

27. offerFirst(E e)

将指定的数据添加到此集合的头部，返回 boolean 值。接收 E 入参，作为指定数据。

28. offerLast(E e)

将指定的数据添加到此集合的尾部，返回 boolean 值。接收 E 入参，作为指定数据。

29. peekFirst()

返回此集合头部位置的数据，如果数据不存在，则返回空。

30. peekLast()

返回此集合尾部位置的数据，如果数据不存在，则返回空。

31. pollFirst()

返回此集合头部位置的数据，并删除此数据。如果数据不存在，则返回空，代码如下：

```java
//第 14 章/three/LinkedListTest.java
public class LinkedListTest {

    public static void main(String[] args) {
```

```
        LinkedList<Integer> linkedList = new LinkedList<>();
        linkedList.add(18);
        linkedList.add(1545);
        linkedList.add(455);
        linkedList.add(17848);
        linkedList.add(655);
        System.out.println(linkedList);
        System.out.println(linkedList.pollFirst());
        System.out.println(linkedList);
    }

}
```

执行结果如下：

```
[18, 1545, 455, 17848, 655]
18
[1545, 455, 17848, 655]
```

32. pollLast()

返回此集合尾部位置的数据，并删除此数据。如果数据不存在，则返回空。

33. pop()

返回此集合头部位置的数据，并删除此数据。如果数据不存在，则抛出异常。

34. removeFirst()

返回此集合头部位置的数据，并删除此数据。如果数据不存在，则抛出异常。

35. removeLast()

返回此集合尾部位置的数据，并删除此数据。如果数据不存在，则抛出异常。

14.4 Vector 类

此类实现了 List 接口，底层基于数组实现，具有多线程并发安全、自动扩容、允许重复数据、允许空数据的特性。

14.4.1 构造器

Vector 类构造器，见表 14-3。

表 14-3 Vector 类构造器

构 造 器	描 述
Vector()	构造新的对象，默认为无参构造器
Vector(int initialCapacity)	构造新的对象，指定初始容量

续表

构 造 器	描 述
Vector(int initialCapacity, int capacityIncrement)	构造新的对象,指定初始容量,指定扩容容量
Vector(Collection <? extends E > c)	构造新的对象,指定初始数据

14.4.2 常用方法

本节基于官方文档介绍常用方法,核心方法配合代码示例以方便读者理解。

1. add(int index, E element)

将指定的数据插入此集合指定的位置。接收 int 入参,作为指定位置,接收 E 入参,作为指定数据,代码如下:

```
//第 14 章/four/VectorTest.java
public class VectorTest {

    public static void main(String[] args) {
        Vector < Integer > vector = new Vector <>();
        vector.add(18);
        vector.add(3223);
        vector.add(23423);
        vector.add(1234238);
        vector.add(4234);
        vector.add(2,88888);
        System.out.println(vector);
    }
}
```

执行结果如下:

```
[18, 3223, 88888, 23423, 1234238, 4234]
```

2. add(E e)

将指定的数据添加到此集合的尾部,返回 boolean 值。接收 E 入参,作为指定数据。

3. addAll(int index, Collection <? extends E > c)

将指定的集合数据插入此集合指定的位置,返回 boolean 值。接收 int 入参,作为指定位置,接收 Collection 入参,作为指定集合数据。

4. addAll(Collection <? extends E > c)

将指定的集合数据添加到此集合的尾部,返回 boolean 值。接收 Collection 入参,作为指定集合数据。

5. clear()

清空此集合内的所有数据。

6. contains(Object o)

检查此集合中是否包含指定的数据，返回 boolean 值。接收 Object 入参，作为指定数据。

7. containsAll(Collection <?> c)

检查此集合中是否包含指定的集合数据，返回 boolean 值。接收 Collection 入参，作为指定集合数据。

8. equals(Object o)

对此对象与指定对象进行比较，返回 boolean 值。接收 Object 入参，作为指定对象。

9. get(int index)

返回此集合中指定位置的数据。接收 int 入参，作为指定位置。

10. indexOf(Object o)

返回此集合中首次出现的指定数据的索引，如果此集合不包含该数据，则返回－1。

11. isEmpty()

检查此集合是否为空，返回 boolean 值。

12. lastIndexOf(Object o)

返回此集合中最后一次出现的指定数据的索引，如果此集合不包含该数据，则返回－1。

13. remove(int index)

删除此集合中指定位置的数据，并返回删除的数据。接收 int 入参，作为指定位置。

14. remove(Object o)

从此集合中删除首次出现的指定数据，返回 boolean 值。接收 Object 入参，作为指定数据。

15. removeAll(Collection <?> c)

从此集合中删除指定集合中包含的所有数据，返回 boolean 值。接收 Collection 入参，作为指定集合数据。

16. replaceAll(UnaryOperator <E> operator)

根据指定的数据规则从此集合中替换所有数据。接收 UnaryOperator 入参，作为指定数据规则。

17. set(int index, E element)

将此集合中指定位置的数据替换为指定的数据，并返回原数据。接收 int 入参，作为指定位置，接收 E 入参，作为指定数据。

18. size()

返回此集合中数据的长度。

19. sort(Comparator <? super E > c)

根据指定的比较器所引导的顺序对该集合数据进行排序。接收 Comparator 入参，作为指定的比较器。

20. subList(int fromIndex, int toIndex)

从指定的开始位置到指定的结束位置分割此集合数据，并返回新的集合对象。接收 int 入参，作为开始位置，接收 int 入参，作为结束位置。

21. removeIf(Predicate <? super E > filter)

根据指定的条件规则从此集合中删除数据。接收 Predicate 入参，作为指定条件规则。

22. trimToSize()

将此集合容器的容量缩减为此集合数据的长度。

23. addElement(E obj)

将指定的数据添加到此集合的尾部。接收 E 入参，作为指定数据。

24. capacity()

返回此集合中容器的长度，代码如下：

```
//第 14 章/four/VectorTest.java
public class VectorTest {

    public static void main(String[] args) {
        Vector < Integer > vector = new Vector <>(16);
        vector.add(18);
        vector.add(3223);
        vector.add(23423);
        vector.add(1234238);
        vector.add(4234);
        vector.addElement(88888);
        System.out.println(vector);
        System.out.println(vector.size());
        System.out.println(vector.capacity());
    }
}
```

执行结果如下：

```
[18, 3223, 23423, 1234238, 4234, 88888]
6
16
```

25. elementAt(int index)

返回此集合中指定位置的数据。接收 int 入参，作为指定位置，代码如下：

```java
//第 14 章/four/VectorTest.java
public class VectorTest {

    public static void main(String[] args) {
        Vector<Integer> vector = new Vector<>(16);
        vector.add(18);
        vector.add(3223);
        vector.add(23423);
        vector.add(1234238);
        vector.add(4234);
        vector.addElement(88888);
        System.out.println(vector);
        System.out.println(vector.elementAt(3));
        System.out.println(vector.get(3));
    }
}
```

执行结果如下：

```
[18, 3223, 23423, 1234238, 4234, 88888]
1234238
1234238
```

26. firstElement()

返回此集合头部位置的数据。如果数据为空,则抛出异常。

27. lastElement()

返回此集合尾部位置的数据。如果数据为空,则抛出异常。

28. removeElementAt(int index)

删除指定索引处的数据,代码如下：

```java
//第 14 章/four/VectorTest.java
public class VectorTest {
    public static void main(String[] args) {
        Vector<Integer> vector = new Vector<>();
        vector.add(18);
        vector.add(3223);
        vector.add(23423);
        vector.add(1234238);
        vector.add(4234);
        vector.add(2,88888);
        System.out.println(vector);
        vector.removeElementAt(2);
        System.out.println(vector);
    }
}
```

执行结果如下：

```
[18, 3223, 88888, 23423, 1234238, 4234]
[18, 3223, 23423, 1234238, 4234]
```

14.5 Iterator 接口

Iterator 接口定义了迭代器的相关规范,增强 for 循环也是基于此接口实现的。本节基于官方文档介绍常用方法,核心方法配合代码示例以方便读者理解。

1. hasNext()

检查此迭代器中是否还有数据,返回 boolean 值。

2. next()

返回此迭代器当前的数据,代码如下:

```java
//第 14 章/five/IteratorTest.java
public class IteratorTest {
    public static void main(String[] args) {
        Vector<Integer> vector = new Vector<>();
        vector.add(18);
        vector.add(19);
        vector.add(20);
        vector.add(21);
        vector.add(22);
        vector.add(23);
        vector.add(24);
        vector.add(25);
        vector.add(26);

        Iterator<Integer> iterator = vector.iterator();
        while (iterator.hasNext()) {
            Integer next = iterator.next();
            System.out.println(next);
        }
    }

}
```

执行结果如下:

```
18
19
20
21
22
23
24
25
26
```

在使用此迭代器的过程中，不允许通过源数据进行新增、删除操作，代码如下：

```
//第 14 章/five/IteratorTest.java
public class IteratorTest {
    public static void main(String[] args) {
        Vector<Integer> vector = new Vector<>();
        vector.add(18);
        vector.add(19);
        vector.add(20);
        vector.add(21);
        vector.add(22);
        vector.add(23);
        vector.add(24);
        vector.add(25);
        vector.add(26);

        Iterator<Integer> iterator = vector.iterator();
        while (iterator.hasNext()) {
            Integer next = iterator.next();
            System.out.println(next);
            vector.add(6666);
//vector.remove(3); 错误使用
        }
    }

}
```

执行结果如下：

```
18
Exception in thread "main" java.util.ConcurrentModificationException
    at java.base/java.util.Vector$Itr.
    checkForComodification(Vector.java:1298)
    at java.base/java.util.Vector$Itr.next(Vector.java:1254)
    at cn.kungreat.list.IteratorTest.main(IteratorTest.java:21)
```

3. remove()

从数据集合中删除此迭代器当前的数据，代码如下：

```
//第 14 章/five/IteratorTest.java
public class IteratorTest {
    public static void main(String[] args) {
        Vector<Integer> vector = new Vector<>();
        vector.add(18);
        vector.add(19);
        vector.add(20);
        vector.add(21);
        vector.add(22);
        vector.add(23);
        vector.add(24);
```

```
        vector.add(25);
        vector.add(26);

        Iterator<Integer> iterator = vector.iterator();
        while (iterator.hasNext()) {
            Integer next = iterator.next();
            System.out.println(next);
            if(next.equals(22)){
                iterator.remove();
            }
        }
        System.out.println(vector);
    }

}
```

执行结果如下：

```
18
19
20
21
22
23
24
25
26
[18, 19, 20, 21, 23, 24, 25, 26]
```

使用增强 for 遍历数据，代码如下：

```
//第 14 章/five/IteratorTest.java
public class IteratorTest {

    public static void main(String[] args) {
        Vector<Integer> vector = new Vector<>();
        vector.add(18);
        vector.add(19);
        vector.add(20);
        vector.add(21);
        vector.add(22);
        vector.add(23);
        vector.add(24);
        vector.add(25);
        vector.add(26);
        for (Integer num : vector) {
            System.out.println(num);
        }
    }

}
```

执行结果如下：

```
18
19
20
21
22
23
24
25
26
```

8min

14.6 ListIterator 接口

此接口继承了 Iterator 接口并扩展了一些方法。可以在迭代的过程中增加数据。本节基于官方文档介绍常用方法，核心方法配合代码示例以方便读者理解。

1. hasNext()

检查此迭代器中是否还有数据，返回 boolean 值。

2. next()

返回此迭代器当前的数据。

3. add(E e)

在此迭代器当前的位置中插入新数据。接收 E 入参，作为插入数据，代码如下：

```java
//第 14 章/six/ListIteratorTest.java
public class ListIteratorTest {

    public static void main(String[] args) {
        Vector<Integer> vector = new Vector<>();
        vector.add(180);
        vector.add(190);
        vector.add(200);
        vector.add(210);
        vector.add(220);
        vector.add(230);
        vector.add(240);
        vector.add(250);
        vector.add(260);
        ListIterator<Integer> listIterator = vector.listIterator();
        while (listIterator.hasNext()) {
            Integer next = listIterator.next();
            System.out.println(next);
            if(next.equals(200)){
                listIterator.add(99999999);
            }
        }
```

```
        System.out.println(vector);
    }
}
```

执行结果如下：

```
180
190
200
210
220
230
240
250
260
[180, 190, 200, 99999999, 210, 220, 230, 240, 250, 260]
```

4. hasPrevious()

检查此迭代器中是否还有数据(反向遍历)，返回 boolean 值。

5. previous()

返回此迭代器当前的数据(反向遍历)，代码如下：

```
//第 14 章/six/ListIteratorTest.java
public class ListIteratorTest {

    public static void main(String[] args) {
        Vector<Integer> vector = new Vector<>();
        vector.add(180);
        vector.add(190);
        vector.add(200);
        vector.add(210);
        vector.add(220);
        vector.add(230);
        vector.add(240);
        vector.add(250);
        vector.add(260);
        ListIterator<Integer> listIterator =
                                    vector.listIterator(vector.size());
        while (listIterator.hasPrevious()) {
            Integer previous = listIterator.previous();
            System.out.println(previous);
        }
    }
}
```

执行结果如下：

```
260
250
```

```
240
230
220
210
200
190
180
```

6. nextIndex()

返回当前索引的位置，如图 14-3 所示。

```
//Vector类
final class ListItr extends Itr implements ListIterator<E> {
    ListItr(int index) {
        super();
        cursor = index;
    }

    public boolean hasPrevious() {
        return cursor != 0;
    }

    public int nextIndex() {
        return cursor;
    }

    public int previousIndex() {
        return cursor - 1;
    }
}
```

图 14-3 源代码

7. previousIndex()

返回上一个索引的位置。

小结

本章介绍了 List 接口的常用实现类，读者需要理解各实现类之间的差异。

习题

1. 判断题

(1) ArrayList 类底层基于数组实现。(　　)

(2) LinkedList 类底层基于链表实现。(　　)

(3) 数组实现类的查、改功能往往比链表实现类的相关功能更快。(　　)

(4) 数组实现类的扩容功能往往比链表实现类的相关功能更慢。(　　)

(5) Vector 类底层基于数组实现。(　　)

2. 选择题

(1) 返回 List 集合数据长度的方法是(　　)。(单选)

A. isEmpty()　　B. iterator()

C. size()　　D. toArray()

(2) List 集合将指定的数据添加到此集合尾部的方法是(　　)。(单选)

A. isEmpty()　　B. iterator()

C. size()　　D. add(E e)

(3) List 集合将指定的数据插入此集合中指定位置的方法是(　　)。(单选)

A. add(int index, E element)　　B. iterator()

C. size()　　D. add(E e)

(4) 检查 List 集合中是否包含指定数据的方法是(　　)。(单选)

A. add(int index, E element)　　B. contains(Object o)

C. size()　　D. add(E e)

(5) 返回 List 集合中指定位置数据的方法是(　　)。(单选)

A. add(int index, E element)　　B. contains(Object o)

C. get(int index)　　D. add(E e)

3. 填空题

查看执行结果并补充代码,代码如下:

```
//第 14 章/answer/OneAnswer.java
public class OneAnswer {

    public static void main(String[] args) {
        ArrayList<Integer> arrayList = new ArrayList<>(16);
        arrayList.add(15);
        arrayList.add(16);
        arrayList.add(17);
        arrayList.add(18);
        System.out.println(arrayList);
        arrayList.________________;
        System.out.println(arrayList);
        arrayList.________________;
        System.out.println(arrayList);
    }
}
```

执行结果如下:

```
[15, 16, 17, 18]
[15, 16, 886, 17, 18]
[]
```

第 15 章
CHAPTER 15

Set 集合框架

Set 接口继承了 Collection 接口，Set 接口被称为无序集合框架。不包含重复的元素，需要注意元素的 hashCode()、equals(Object obj)方法。

15.1 Set 接口

Set 接口继承了 Collection 接口，定义了 Set 接口的相关规范方法。本节基于官方文档介绍常用方法，核心方法配合代码示例以方便读者理解。

1. of(E... elements)

静态方法，从指定的数据中创建此集合对象。接收 E 泛型入参，作为指定数据，代码如下：

```
//第 15 章/one/SetTest.java
public class SetTest {

    public static void main(String[] args) {
        Set<Integer> integerSet = Set.of(15, 454, 4656, 65, 78, 96, 35,
                                        24, 54, 87, 545);
        System.out.println(integerSet);
    }
}
```

大致的执行结果如下：

```
[24, 96, 54, 78, 35, 454, 15, 4656, 545, 65, 87]
```

此方法将创建一个默认的 ImmutableCollections.SetN 实现类，如图 15-1 所示。

此实现类继承了 AbstractImmutableCollection 类，查看源代码发现此类不允许新增、删除数据，如图 15-2 所示。

2. add(E e)

如果指定的数据不存在，则将指定的数据添加至此集合中，返回 boolean 值。接收 E 泛型入参，作为指定数据。

```
@SafeVarargs
@SuppressWarnings("varargs")
static <E> Set<E> of(E... elements) {
    switch (elements.length) {
        case 0:
            @SuppressWarnings("unchecked")
            var set = (Set<E>) ImmutableCollections.EMPTY_SET;
            return set;
            case 1:
                return new ImmutableCollections.Set12<>(elements[0]);
                case 2:
                    return new ImmutableCollections.Set12<>(elements[0], elements[1]);
                    default:
                        return new ImmutableCollections.SetN<>(elements);
    }
}
```

图 15-1 of(E... elements)源代码

```
@jdk.internal.ValueBased
static abstract class AbstractImmutableCollection<E> extends AbstractCollection<E> {
    //所有方法都抛出UnsupportedOperationException
    @Override public boolean add(E e) { throw uoe(); }
    @Override public boolean addAll(Collection<? extends E> c) { throw uoe(); }
    @Override public void    clear() { throw uoe(); }
    @Override public boolean remove(Object o) { throw uoe(); }
    @Override public boolean removeAll(Collection<?> c) { throw uoe(); }
    @Override public boolean removeIf(Predicate<? super E> filter) { throw uoe(); }
    @Override public boolean retainAll(Collection<?> c) { throw uoe(); }
}
```

图 15-2 AbstractImmutableCollection 类的源代码

3. addAll(Collection <? extends E > c)

如果指定的数据不存在,则将指定的数据添加至此集合中,返回 boolean 值。接收 Collection 入参,作为指定数据集合。

4. clear()

清空此集合中的所有数据。

5. contains(Object o)

检查此集合中是否包含指定的数据,返回 boolean 值。接收 Object 入参,作为指定数据。

6. containsAll(Collection <?> c)

检查此集合中是否包含指定的集合数据,返回 boolean 值。接收 Collection 入参,作为指定集合数据。

7. equals(Object o)

对此对象与指定对象进行比较,返回 boolean 值。接收 Object 入参,作为指定对象。

8. isEmpty()

检查此集合是否为空,返回 boolean 值。

9. iterator()

返回此集合中数据的迭代器，代码如下：

```
//第 15 章/one/SetTest.java
public class SetTest {

    public static void main(String[] args) {
        Set<Integer> integerSet = Set.of(15, 454, 4656, 65, 78, 96, 35,
                                        24, 54, 87, 545);
        Iterator<Integer> iterator = integerSet.iterator();
        while (iterator.hasNext()) {
            Integer next = iterator.next();
            System.out.println(next);
        }
    }
}
```

大致的执行结果如下：

```
24
87
65
545
4656
15
454
35
78
54
96
```

此默认的迭代器实现类是 SetNIterator 类，此类的迭代顺序有一定的随机性，如图 15-3 所示。

```
private final class SetNIterator implements Iterator<E> {

    private int remaining;

    private int idx;

    SetNIterator() {
        remaining = size;
        idx = (int) ((SALT32L * elements.length) >>> 32);
    }
    //省略...
}
```

图 15-3 SetNIterator 类的源代码

10. remove(Object o)

从此集合中删除指定的数据，返回 boolean 值。接收 Object 入参，作为指定数据。

11. removeAll(Collection <? > c)

从此集合中删除指定的数据，返回 boolean 值。接收 Collection 入参，作为指定集合数据。

12. retainAll(Collection <? > c)

从此集合中删除未排除的数据，返回 boolean 值。接收 Collection 入参，作为排除的集合数据。

13. size()

返回此集合中数据的长度。

14. removeIf(Predicate <? super E > filter)

根据指定的条件规则从此集合中删除数据。接收 Predicate 入参，作为指定条件规则。

15.2 HashSet 类

HashSet 类实现了 Set 接口。不允许重复的数据、无序列表、自动扩容机制，允许空元素。底层包装了一个 HashMap 实现类。

15.2.1 构造器

HashSet 类构造器，见表 15-1。

表 15-1　HashSet 类构造器

构　造　器	描　　述
HashSet()	构造新的对象，默认为无参构造器
HashSet(int initialCapacity)	构造新的对象，指定初始容量
HashSet(int initialCapacity, float loadFactor)	构造新的对象，指定初始容量，指定负载因子
HashSet(Collection <? extends E > c)	构造新的对象，指定初始数据

15.2.2 常用方法

本节基于官方文档介绍常用方法，核心方法配合代码示例以方便读者理解。

1. add(E e)

如果指定的数据不存在，则将指定的数据添加至此集合中，返回 boolean 值。接收 E 入参，作为指定数据，代码如下：

```
//第 15 章/two/HashSetTest.java
public class HashSetTest {

    public static void main(String[] args) {
```

```
        HashSet < Integer > hashSet = new HashSet <>();
        hashSet.add(5656);
        hashSet.add(454);
        hashSet.add(235);
        hashSet.add(564);
        hashSet.add(6456);
        hashSet.add(456);
        hashSet.add(245);
        System.out.println(hashSet.add(5656));
        System.out.println(hashSet);
    }
}
```

执行结果如下：

```
false
[564, 245, 454, 5656, 6456, 456, 235]
```

2. addAll(Collection <? extends E > c)

如果指定的数据不存在，则将指定的数据添加至此集合中，返回 boolean 值。接收 Collection 入参，作为指定数据集合。

3. clear()

清空此集合中的所有数据。

4. contains(Object o)

检查此集合中是否包含指定的数据，返回 boolean 值。接收 Object 入参，作为指定数据，代码如下：

```
//第 15 章/two/HashSetTest.java
public class HashSetTest {

    public static void main(String[] args) {
        HashSet < Integer > hashSet = new HashSet <>();
        hashSet.add(5656);
        hashSet.add(454);
        hashSet.add(235);
        hashSet.add(564);
        hashSet.add(6456);
        hashSet.add(456);
        hashSet.add(245);
        System.out.println(hashSet.add(5656));
        System.out.println(hashSet.contains(235));
    }
}
```

执行结果如下：

```
false
true
```

5. isEmpty()

检查此集合是否为空，返回 boolean 值。

6. remove(Object o)

从此集合中删除指定的数据，返回 boolean 值。接收 Object 入参，作为指定数据，代码如下：

```
//第 15 章/two/HashSetTest.java
public class HashSetTest {

    public static void main(String[] args) {
        HashSet<Integer> hashSet = new HashSet<>();
        hashSet.add(5656);
        hashSet.add(454);
        hashSet.add(235);
        hashSet.add(564);
        hashSet.add(6456);
        hashSet.add(456);
        hashSet.add(245);
        System.out.println(hashSet.remove(5656));
        System.out.println(hashSet);
    }
}
```

执行结果如下：

```
true
[564, 245, 454, 6456, 456, 235]
```

7. size()

返回此集合中数据的长度。

8. iterator()

返回此集合中数据的迭代器。

9. removeAll(Collection<?> c)

从此集合中删除指定的数据，返回 boolean 值。接收 Collection 入参，作为指定集合数据。

10. retainAll(Collection<?> c)

从此集合中删除未排除的数据，返回 boolean 值。接收 Collection 入参，作为排除的集合数据，代码如下：

```java
//第15章/two/HashSetTest.java
public class HashSetTest {

    public static void main(String[] args) {
        HashSet<Integer> hashSet = new HashSet<>();
        hashSet.add(5656);
        hashSet.add(454);
        hashSet.add(235);
        hashSet.add(564);
        hashSet.add(6456);
        hashSet.add(456);
        hashSet.add(245);
        System.out.println(hashSet);
        hashSet.retainAll(Set.of(235,564,245));
        System.out.println(hashSet);
    }
}
```

执行结果如下：

```
[564, 245, 454, 5656, 6456, 456, 235]
[564, 245, 235]
```

11. removeIf(Predicate<? super E> filter)

根据指定的条件规则从此集合中删除数据。接收 Predicate 入参，作为指定条件规则，代码如下：

```java
//第15章/two/HashSetTest.java
public class HashSetTest {

    public static void main(String[] args) {
        HashSet<Integer> hashSet = new HashSet<>();
        hashSet.add(5656);
        hashSet.add(454);
        hashSet.add(235);
        hashSet.add(564);
        hashSet.add(6456);
        hashSet.add(456);
        hashSet.add(245);
        System.out.println(hashSet);
        hashSet.removeIf(new Predicate<Integer>() {
            @Override
            public boolean test(Integer integer) {
                return integer.equals(454);
            }
        });
        System.out.println(hashSet);
    }
}
```

执行结果如下：

```
[564, 245, 454, 5656, 6456, 456, 235]
[564, 245, 5656, 6456, 456, 235]
```

12. equals(Object o)

对此对象与指定对象进行比较，返回 boolean 值。接收 Object 入参，作为指定对象，代码如下：

```
//第 15 章/two/HashSetTest.java
public class HashSetTest {

    public static void main(String[] args) {
        HashSet<Integer> hashSet = new HashSet<>();
        hashSet.add(5656);
        hashSet.add(454);
        hashSet.add(235);
        hashSet.add(564);
        hashSet.add(6456);
        System.out.println(hashSet);
        System.out.println(hashSet.equals(Set.of(454,235,564,6456,5656)));
    }
}
```

执行结果如下：

```
[564, 454, 5656, 6456, 235]
true
```

集合数据的 equals(Object o)方法，比较的是集合中的所有数据类型是否相同、长度是否相等、数据是否相等，如图 15-4 所示。

```
public boolean equals(Object o) {
    if (o == this)
        return true;

    if (!(o instanceof Set))
        return false;
    Collection<?> c = (Collection<?>) o;
    if (c.size() != size())
        return false;
    try {
        return containsAll(c);
    } catch (ClassCastException | NullPointerException unused) {
        return false;
    }
}
```

图 15-4 equals(Object o)源代码

15.3 LinkedHashSet 类

12min

LinkedHashSet 类继承了 HashSet 类实现了 Set 接口。不允许重复的数据、有序链表，允许空元素。底层包装了一个 LinkedHashMap 实现类。

15.3.1 构造器

LinkedHashSet 类构造器，见表 15-2。

表 15-2 LinkedHashSet 类构造器

构 造 器	描 述
LinkedHashSet()	构造新的对象，默认为无参构造器
LinkedHashSet(int initialCapacity)	构造新的对象，指定初始容量
LinkedHashSet(int initialCapacity, float loadFactor)	构造新的对象，指定初始容量，指定负载因子
LinkedHashSet(Collection <? extends E > c)	构造新的对象，指定初始数据

15.3.2 常用方法

本节基于官方文档介绍常用方法，核心方法配合代码示例以方便读者理解。

1. add(E e)

如果指定的数据不存在，则将指定的数据添加至此集合中，返回 boolean 值。接收 E 入参，作为指定数据，代码如下：

```
//第 15 章/three/LinkedHashSetTest.java
public class LinkedHashSetTest {

    public static void main(String[] args) {
        LinkedHashSet<Integer> linkedHashSet = new LinkedHashSet<>();
        linkedHashSet.add(455);
        linkedHashSet.add(232);
        linkedHashSet.add(23214);
        linkedHashSet.add(4534);
        linkedHashSet.add(6767);
        linkedHashSet.add(454);
        linkedHashSet.add(5454);
        linkedHashSet.add(455);
        System.out.println(linkedHashSet);
    }
}
```

执行结果如下：

```
[455, 232, 23214, 4534, 6767, 454, 5454]
```

2. addAll(Collection <? extends E > c)

如果指定的数据不存在，则将指定的数据添加至此集合中，返回 boolean 值。接收 Collection 入参，作为指定数据集合。

3. clear()

清空此集合中的所有数据。

4. contains(Object o)

检查此集合中是否包含指定的数据，返回 boolean 值。接收 Object 入参，作为指定数据。

5. isEmpty()

检查此集合是否为空，返回 boolean 值。

6. remove(Object o)

从此集合中删除指定的数据，返回 boolean 值。接收 Object 入参，作为指定数据，代码如下：

```
//第 15 章/three/LinkedHashSetTest.java
public class LinkedHashSetTest {

    public static void main(String[] args) {
        LinkedHashSet<Integer> linkedHashSet = new LinkedHashSet<>();
        linkedHashSet.add(455);
        linkedHashSet.add(232);
        linkedHashSet.add(23214);
        linkedHashSet.add(4534);
        linkedHashSet.add(6767);
        linkedHashSet.add(454);
        linkedHashSet.add(5454);
        System.out.println(linkedHashSet);
        System.out.println(linkedHashSet.remove(4534));
        System.out.println(linkedHashSet);
    }
}
```

执行结果如下：

```
[455, 232, 23214, 4534, 6767, 454, 5454]
true
[455, 232, 23214, 6767, 454, 5454]
```

7. size()

返回此集合中数据的长度。

8. iterator()

返回此集合中数据的迭代器。

9. removeAll(Collection<?> c)

从此集合中删除指定的数据，返回 boolean 值。接收 Collection 入参，作为指定集合数据，代码如下：

```
//第 15 章/three/LinkedHashSetTest.java
public class LinkedHashSetTest {
```

```
    public static void main(String[] args) {
        LinkedHashSet<Integer> linkedHashSet = new LinkedHashSet<>();
        linkedHashSet.add(455);
        linkedHashSet.add(232);
        linkedHashSet.add(23214);
        linkedHashSet.add(4534);
        linkedHashSet.add(6767);
        linkedHashSet.add(454);
        linkedHashSet.add(5454);
        System.out.println(linkedHashSet);
        System.out.println(linkedHashSet.removeAll(List.of(232, 23214)));
        System.out.println(linkedHashSet);
    }
}
```

执行结果如下：

```
[455, 232, 23214, 4534, 6767, 454, 5454]
true
[455, 4534, 6767, 454, 5454]
```

10. retainAll(Collection<?> c)

从此集合中删除未排除的数据，返回 boolean 值。接收 Collection 入参，作为排除的集合数据。

11. removeIf(Predicate<? super E> filter)

根据指定的条件规则从此集合中删除数据。接收 Predicate 入参，作为指定条件规则。

12. equals(Object o)

对此对象与指定对象进行比较，返回 boolean 值。接收 Object 入参，作为指定对象。

15.4 TreeSet 类

TreeSet 类实现了 NavigableSet 接口。不允许重复的数据、有排序规则的树结构、不允许空元素。底层包装了一个 TreeMap 实现类。

15.4.1 构造器

TreeSet 类构造器，见表 15-3。

表 15-3 TreeSet 类构造器

构 造 器	描 述
TreeSet()	构造新的对象，默认为无参构造器
TreeSet(Collection<? extends E> c)	构造新的对象，指定初始数据

续表

构 造 器	描 述
TreeSet(Comparator <? super E > comparator)	构造新的对象,指定公共比较器
TreeSet(SortedSet < E > s)	构造新的对象,指定初始数据并指定公共比较器

15.4.2 常用方法

本节基于官方文档介绍常用方法,核心方法配合代码示例以方便读者理解。

1. add(E e)

如果指定的数据不存在,则将指定的数据添加至此集合中,返回 boolean 值。接收 E 入参,作为指定数据,代码如下:

```
//第 15 章/four/TreeSetTest.java
public class TreeSetTest {

    public static void main(String[] args) {
        TreeSet < Integer > treeSet = new TreeSet <>();
        treeSet.add(4545);
        treeSet.add(232);
        treeSet.add(4534);
        treeSet.add(45345);
        treeSet.add(45);
        treeSet.add(432);
        treeSet.add(78);
        treeSet.add(123);
        treeSet.add(232);
        System.out.println(treeSet);
    }
}
```

执行结果如下:

```
[45, 78, 123, 232, 432, 4534, 4545, 45345]
```

2. addAll(Collection <? extends E > c)

如果指定的数据不存在,则将指定的数据添加至此集合中,返回 boolean 值。接收 Collection 入参,作为指定数据集合。

3. clear()

清空此集合中的所有数据。

4. contains(Object o)

检查此集合中是否包含指定的数据,返回 boolean 值。接收 Object 入参,作为指定数据。

5．isEmpty()

检查此集合是否为空，返回 boolean 值。

6．remove(Object o)

从此集合中删除指定的数据，返回 boolean 值。接收 Object 入参，作为指定数据。

7．size()

返回此集合中数据的长度，代码如下：

```
//第 15 章/four/TreeSetTest.java
public class TreeSetTest {

    public static void main(String[] args) {
        TreeSet<Integer> treeSet = new TreeSet<>();
        treeSet.add(4545);
        treeSet.add(232);
        treeSet.add(4534);
        treeSet.add(45345);
        treeSet.add(45);
        treeSet.add(432);
        treeSet.add(78);
        treeSet.add(123);
        treeSet.add(232);
        System.out.println(treeSet);
        System.out.println(treeSet.size());
    }
}
```

执行结果如下：

```
[45, 78, 123, 232, 432, 4534, 4545, 45345]
8
```

8．iterator()

返回此集合中数据的迭代器。

9．removeAll(Collection <? > c)

从此集合中删除指定的数据，返回 boolean 值。接收 Collection 入参，作为指定集合数据。

10．retainAll(Collection <? > c)

从此集合中删除未排除的数据，返回 boolean 值。接收 Collection 入参，作为排除的集合数据。

11．removeIf(Predicate <? super E > filter)

根据指定的条件规则从此集合中删除数据。接收 Predicate 入参，作为指定条件规则。

12. equals(Object o)

对此对象与指定对象进行比较,返回 boolean 值。接收 Object 入参,作为指定对象。

13. ceiling(E e)

返回此集合中大于或等于指定数据的最小元素,如果没有此元素,则返回空。接收 E 入参,作为指定数据,代码如下:

```
//第 15 章/four/TreeSetTest.java
public class TreeSetTest {

    public static void main(String[] args) {
        TreeSet < Integer > treeSet = new TreeSet <>();
        treeSet.add(4545);
        treeSet.add(232);
        treeSet.add(4534);
        treeSet.add(45345);
        treeSet.add(45);
        treeSet.add(432);
        treeSet.add(78);
        treeSet.add(123);
        System.out.println(treeSet);
        System.out.println(treeSet.ceiling(231));
    }
}
```

执行结果如下:

```
[45, 78, 123, 232, 432, 4534, 4545, 45345]
232
```

14. floor(E e)

返回此集合中小于或等于指定数据的最大元素,如果没有此元素,则返回空。接收 E 入参,作为指定数据。

15. first()

返回当前在此集合中的第 1 个元素,代码如下:

```
//第 15 章/four/TreeSetTest.java
public class TreeSetTest {

    public static void main(String[] args) {
        TreeSet < Integer > treeSet = new TreeSet <>();
        treeSet.add(4545);
        treeSet.add(232);
        treeSet.add(4534);
        treeSet.add(45345);
        treeSet.add(45);
        treeSet.add(432);
```

```
        treeSet.add(78);
        treeSet.add(123);
        System.out.println(treeSet.first());
        System.out.println(treeSet);
    }
}
```

执行结果如下:

```
45
[45, 78, 123, 232, 432, 4534, 4545, 45345]
```

16. last()

返回当前在此集合中的最后一个元素。

17. pollFirst()

返回当前在此集合中的第1个元素并删除此元素。如果此集合为空,则返回空,代码如下:

```
//第15章/four/TreeSetTest.java
public class TreeSetTest {

    public static void main(String[] args) {
        TreeSet<Integer> treeSet = new TreeSet<>();
        treeSet.add(4545);
        treeSet.add(232);
        treeSet.add(4534);
        treeSet.add(45345);
        treeSet.add(45);
        treeSet.add(432);
        treeSet.add(78);
        treeSet.add(123);
        System.out.println(treeSet.pollFirst());
        System.out.println(treeSet);
    }
}
```

执行结果如下:

```
45
[78, 123, 232, 432, 4534, 4545, 45345]
```

18. pollLast()

返回当前在此集合中的最后一个元素并删除此元素。如果此集合为空,则返回空。

19. headSet(E toElement, boolean inclusive)

返回此集合的子元素视图,其元素小于或等于指定的数据。接收E入参,作为指定数据,接收boolean入参,作为是否包含相等数据,代码如下:

```java
//第 15 章/four/TreeSetTest.java
public class TreeSetTest {

    public static void main(String[] args) {
        TreeSet<Integer> treeSet = new TreeSet<>();
        treeSet.add(4545);
        treeSet.add(232);
        treeSet.add(4534);
        treeSet.add(45345);
        treeSet.add(45);
        treeSet.add(432);
        treeSet.add(78);
        treeSet.add(123);
        System.out.println(treeSet.headSet(4545,false));
        System.out.println(treeSet);
    }
}
```

执行结果如下：

```
[45, 78, 123, 232, 432, 4534]
[45, 78, 123, 232, 432, 4534, 4545, 45345]
```

20. tailSet(E fromElement, boolean inclusive)

返回此集合的子元素视图，其元素大于或等于指定的数据。接收 E 入参，作为指定数据，接收 boolean 入参，作为是否包含相等数据，代码如下：

```java
//第 15 章/four/TreeSetTest.java
public class TreeSetTest {

    public static void main(String[] args) {
        TreeSet<Integer> treeSet = new TreeSet<>();
        treeSet.add(4545);
        treeSet.add(232);
        treeSet.add(4534);
        treeSet.add(45345);
        treeSet.add(45);
        treeSet.add(432);
        treeSet.add(78);
        treeSet.add(123);
        System.out.println(treeSet.tailSet(4545,true));
        System.out.println(treeSet);
    }
}
```

执行结果如下：

```
[4545, 45345]
[45, 78, 123, 232, 432, 4534, 4545, 45345]
```

21. subSet(E fromElement, boolean fromInclusive, E toElement, boolean toInclusive)

返回此集合中指定范围内的子元素视图。接收 E 入参，作为开始数据，接收 boolean 入参，作为是否包含开始数据，接收 E 入参，作为结束数据，接收 boolean 入参，作为是否包含结束数据，代码如下：

```
//第 15 章/four/TreeSetTest.java
public class TreeSetTest {

    public static void main(String[] args) {
        TreeSet<Integer> treeSet = new TreeSet<>();
        treeSet.add(4545);
        treeSet.add(232);
        treeSet.add(4534);
        treeSet.add(45345);
        treeSet.add(45);
        treeSet.add(432);
        treeSet.add(78);
        treeSet.add(123);
        System.out.println(treeSet.subSet(232,false,4545,true));
        System.out.println(treeSet);
    }
}
```

执行结果如下：

```
[432, 4534, 4545]
[45, 78, 123, 232, 432, 4534, 4545, 45345]
```

小结

本章介绍了 Set 接口的常用实现类，读者需要理解各实现类之间的差异。

习题

1. 判断题

(1) HashSet 类底层基于数组实现。(　　)

(2) LinkedHashSet 类底层基于链表实现。(　　)

(3) TreeSet 类底层基于数组实现。(　　)

2. 选择题

(1) 返回 Set 集合数据长度的方法是(　　)。(单选)

A. isEmpty()　　B. iterator()
C. size()　　D. toArray()

(2) 检查 Set 集合中是否包含指定数据的方法是(　　)。(单选)

A. add(int index, E element)　　B. contains(Object o)
C. size()　　D. add(E e)

(3) 检查 Set 集合中是否包含指定数据集合的方法是(　　)。(单选)

A. add(int index, E element)　　B. contains(Object o)
C. containsAll(Collection <?> c)　　D. add(E e)

(4) 删除 Set 集合中指定数据的方法是(　　)。(单选)

A. add(int index, E element)　　B. contains(Object o)
C. size()　　D. remove(Object o)

(5) 删除 Set 集合中指定数据集合的方法是(　　)。(单选)

A. removeAll(Collection <?> c)　　B. contains(Object o)
C. size()　　D. remove(Object o)

3. 填空题

查看执行结果并补充代码,代码如下:

```
//第 15 章/answer/OneAnswer.java
public class OneAnswer {

    public static void main(String[] args) {
        HashSet<________> hashSet = new HashSet<>();
        hashSet.add(5656);
        hashSet.add(454);
        hashSet.add(235);
        hashSet.add(564);
        hashSet.add(6456);
        hashSet.add(456);
        hashSet.add(245);
        hashSet.add(5656);
        System.out.println(hashSet);
        System.out.println(hashSet.contains(5656));
        hashSet.________;
        System.out.println(hashSet.contains(5656));
        System.out.println(hashSet);
    }
}
```

执行结果如下:

```
[564, 245, 454, 5656, 6456, 456, 235]
true
false
[564, 245, 454, 6456, 456, 235]
```

第 16 章 CHAPTER 16 Map 集合框架

Map 集合框架存储一对键-值对，其键的作用类似于 Set 接口的功能，其键不允许重复的数据。需要注意键的 hashCode()、equals(Object obj)方法。

22min

16.1 Map 接口

Map 接口定义了相关规范方法。本节基于官方文档介绍常用方法，核心方法配合代码示例以方便读者理解。

1. entry(K k, V v)

静态方法，从指定的数据中创建 Map. Entry 对象。接收 K 入参，作为指定数据的键，接收 V 入参，作为指定数据的值。

2. ofEntries(Map. Entry <? extends K，? extends V >... entries)

静态方法，从指定的数据中创建此集合对象。接收 Map. Entry 入参，作为指定数据，代码如下：

```
//第 16 章/one/MapTest.java
public class MapTest {

    public static void main(String[] args) {
        Map<String, Integer> map = Map.ofEntries(
                Map.entry("one", 4545), Map.entry("two", 4545)
                , Map.entry("three", 4545), Map.entry("four", 4545565)
                , Map.entry("five", 78787), Map.entry("six", 56565));
        System.out.println(map);
    }
}
```

大致的执行结果如下：

```
{four = 4545565, five = 78787, one = 4545, six = 56565, three = 4545, two = 4545}
```

此方法将创建一个默认的 ImmutableCollections. MapN 实现类，如图 16-1 所示。

```java
@SafeVarargs
@SuppressWarnings("varargs")
static <K, V> Map<K, V> ofEntries(Entry<? extends K, ? extends V>... entries) {
    if (entries.length == 0) { //数组的隐式空检查
        @SuppressWarnings("unchecked")
        var map = (Map<K,V>) ImmutableCollections.EMPTY_MAP;
        return map;
    } else if (entries.length == 1) {
        return new ImmutableCollections.Map1<>(entries[0].getKey(),
                                               entries[0].getValue());
    } else {
        Object[] kva = new Object[entries.length << 1];
        int a = 0;
        for (Entry<? extends K, ? extends V> entry : entries) {
            kva[a++] = entry.getKey();
            kva[a++] = entry.getValue();
        }
        return new ImmutableCollections.MapN<>(kva);
    }
}
```

图 16-1 ofEntries 方法的源代码

此实现类继承了 AbstractImmutableMap 类，查看源代码发现此类不允许修改数据，如图 16-2 所示。

```java
@jdk.internal.ValueBased
abstract static class AbstractImmutableMap<K,V> extends AbstractMap<K,V> implements Serializable {
    @Override public void clear() { throw uoe(); }
    @Override public V compute(K key, BiFunction<? super K,? super V,? extends V> rf) { throw uoe(); }
    @Override public V computeIfAbsent(K key, Function<? super K,? extends V> mf) { throw uoe(); }
    @Override public V computeIfPresent(K key, BiFunction<? super K,? super V,? extends V> rf) { throw uoe(); }
    @Override public V merge(K key, V value, BiFunction<? super V,? super V,? extends V> rf) { throw uoe(); }
    @Override public V put(K key, V value) { throw uoe(); }
    @Override public void putAll(Map<? extends K,? extends V> m) { throw uoe(); }
    @Override public V putIfAbsent(K key, V value) { throw uoe(); }
    @Override public V remove(Object key) { throw uoe(); }
    @Override public boolean remove(Object key, Object value) { throw uoe(); }
    @Override public V replace(K key, V value) { throw uoe(); }
    @Override public boolean replace(K key, V oldValue, V newValue) { throw uoe(); }
    @Override public void replaceAll(BiFunction<? super K,? super V,? extends V> f) { throw uoe(); }

    @Override
    public V getOrDefault(Object key, V defaultValue) {
        V v;
        return ((v = get(key)) != null)
                ? v
                : defaultValue;
    }
}
```

图 16-2 AbstractImmutableMap 类的源代码

3. compute(K key, BiFunction <? super K, ? super V, ? extends V > remappingFunction)

根据指定的键计算并处理数据。返回计算函数的结果。接收 K 入参，作为指定的键，接收 BiFunction 入参，作为计算函数。

4. computeIfAbsent(K key, Function <? super K, ? extends V > mappingFunction)

如果指定的键关联的值为空，则根据指定的键计算并处理数据，然后返回计算函数的结果，如果指定的键关联的值非空，则返回指定的键关联的值。接收 K 入参，作为指定的键，接收 Function 入参，作为计算函数。

5. containsKey(Object key)

检查此集合中是否包含指定的键，返回 boolean 值。接收 Object 入参，作为指定的键，代码如下：

```
//第 16 章/one/MapTest.java
public class MapTest {

    public static void main(String[] args) {
        Map<String, Integer> map = Map.ofEntries(
                Map.entry("one", 4545), Map.entry("two", 4545)
                , Map.entry("three", 4545), Map.entry("four", 4545565)
                , Map.entry("five", 78787), Map.entry("six", 56565));
        System.out.println(map);
        System.out.println(map.containsKey("five"));
        System.out.println(map.containsKey("other"));
    }
}
```

大致的执行结果如下：

```
{one=4545, six=56565, three=4545, two=4545, four=4545565, five=78787}
true
false
```

6. containsValue(Object value)

检查此集合中是否包含指定的值，返回 boolean 值。接收 Object 入参，作为指定的值，代码如下：

```
//第 16 章/one/MapTest.java
public class MapTest {

    public static void main(String[] args) {
        Map<String, Integer> map = Map.ofEntries(
                Map.entry("one", 4545), Map.entry("two", 4545)
                , Map.entry("three", 4545), Map.entry("four", 4545565)
                , Map.entry("five", 78787), Map.entry("six", 56565));
        System.out.println(map);
        System.out.println(map.containsValue(123123));
        System.out.println(map.containsValue(4545));
    }
}
```

大致的执行结果如下：

```
{two=4545, three=4545, six=56565, one=4545, five=78787, four=4545565}
false
true
```

7. entrySet()

返回此集合中包含的键、值映射的视图，代码如下：

```
//第 16 章/one/MapTest.java
public class MapTest {

    public static void main(String[] args) {
        Map<String, Integer> map = Map.ofEntries(
                Map.entry("one", 4545), Map.entry("two", 4545)
                , Map.entry("three", 4545), Map.entry("four", 4545565)
                , Map.entry("five", 78787), Map.entry("six", 56565));

        Set<Map.Entry<String, Integer>> entries = map.entrySet();
        for (Map.Entry<String, Integer> entry : entries) {
            System.out.println(entry);
        }
    }
}
```

大致的执行结果如下：

```
six = 56565
one = 4545
five = 78787
four = 4545565
two = 4545
three = 4545
```

ImmutableCollections.MapN 实现类使用的迭代器是 MapNIterator 类，如图 16-3 所示。

```
class MapNIterator implements Iterator<Map.Entry<K,V>> {

    private int remaining;

    private int idx;

    MapNIterator() {
        remaining = size;
        //范围随机基于 SALT32L
        idx = (int) ((SALT32L * (table.length >> 1)) >>> 32) << 1;
    }
}
```

图 16-3 MapNIterator 类的源代码

注意：此迭代器会倒置每次运行主方法，所产生的迭代效果可能会不同。

---

8. equals(Object o)

对此对象与指定对象进行比较，返回 boolean 值。接收 Object 入参，作为指定对象。

9. get(Object key)

返回指定的键所映射的值，如果此集合中不包含指定的键，则返回空。接收 Object 入参，作为指定的键，代码如下：

```
//第 16 章/one/MapTest.java
public class MapTest {

    public static void main(String[] args) {
        Map<String, Integer> map = Map.ofEntries(
                Map.entry("one", 4545), Map.entry("two", 4545)
                , Map.entry("three", 4545), Map.entry("four", 4545565)
                , Map.entry("five", 78787), Map.entry("six", 56565));
        System.out.println(map.get("one"));
        System.out.println(map.get("other"));
    }
}
```

执行结果如下：

```
4545
Null
```

10. getOrDefault(Object key, V defaultValue)

返回指定的键所映射的值，如果此集合中不包含指定的键，则返回默认的值。接收 Object 入参，作为指定的键，接收 V 入参，作为默认的值，代码如下：

```
//第 16 章/one/MapTest.java
public class MapTest {

    public static void main(String[] args) {
        Map<String, Integer> map = Map.ofEntries(
                Map.entry("one", 4545), Map.entry("two", 4545)
                , Map.entry("three", 4545), Map.entry("four", 4545565)
                , Map.entry("five", 78787), Map.entry("six", 56565));
        System.out.println(map.getOrDefault("one",6666));
        System.out.println(map.getOrDefault("other",6666));
    }
}
```

执行结果如下：

```
4545
6666
```

11. isEmpty()

检查此集合是否为空，返回 boolean 值。

12. keySet()

返回此集合中包含键的视图。

13. values()

返回此集合中包含值的视图。

14. put(K key, V value)

将指定的键、值添加到此集合中。返回指定的键相关联的旧值,如果没有,则返回空。接收K入参,作为指定的键,接收V入参,作为指定的值。

15. putAll(Map <? extends K, ? extends V > m)

将指定的集合数据添加到此集合中。接收Map入参,作为指定的集合数据。

16. putIfAbsent(K key, V value)

将指定的键、值添加到此集合中。返回指定的键相关联的旧值,如果没有,则返回空。接收K入参,作为指定的键,接收V入参,作为指定的值。

注意:如果指定的键已经存在,并且其关联的值也不为空,则此方法不会更新值。

---

17. remove(Object key)

从此集合中删除指定的键、值数据,返回指定的键相关联的值,如果没有,则为空。接收Object入参,作为指定的键。

18. remove(Object key, Object value)

仅当指定的键映射到指定的值时,才会从此集合中删除指定的键、值数据,返回boolean值。接收Object入参,作为指定的键,接收Object入参,作为指定的值。

19. replace(K key, V value)

仅当指定的键存在时替换此键所映射的值,返回指定的键相关联的旧值,如果没有,则为空。接收K入参,作为指定的键,接收V入参,作为替换的新值。

20. replace(K key, V oldValue, V newValue)

仅当指定的键存在且映射到指定的值时,才替换此键所映射的值。返回指定的键相关联的旧值,如果没有,则为空。接收K入参,作为指定的键,接收V入参,作为指定键的旧值,接收V入参,作为替换的新值。

21. size()

返回此集合中数据的长度,代码如下:

```
//第 16 章/one/MapTest.java
public class MapTest {

    public static void main(String[] args) {
        Map < String, Integer > map = Map.ofEntries(
                Map.entry("one", 4545), Map.entry("two", 4545)
```

```
            , Map.entry("three", 4545), Map.entry("four", 4545565)
            , Map.entry("five", 78787), Map.entry("six", 56565));
        System.out.println(map.size());
    }
}
```

执行结果如下：

```
6
```

16.2 HashMap 类

HashMap 类底层基于数组和节点实现。数组有自动扩容机制，通过键的 hashCode 值作一些位运算并以此定位到数组内的一个索引点存储数据。当多个数据定位到同一个数组索引时通过节点链接。

16.2.1 数据结构

核心字段如图 16-4 所示。

```
//数据存储
transient Node<K,V>[] table;

//迭代数据
transient Set<Map.Entry<K,V>> entrySet;


//数据长度
transient int size;


//数据修改标记
transient int modCount;

//扩容标记
int threshold;

//负载因子
final float loadFactor;
```

图 16-4 HashMap 类核心字段

数据存储结构简明示意如图 16-5 所示。

16.2.2 构造器

HashMap 类构造器，见表 16-1。

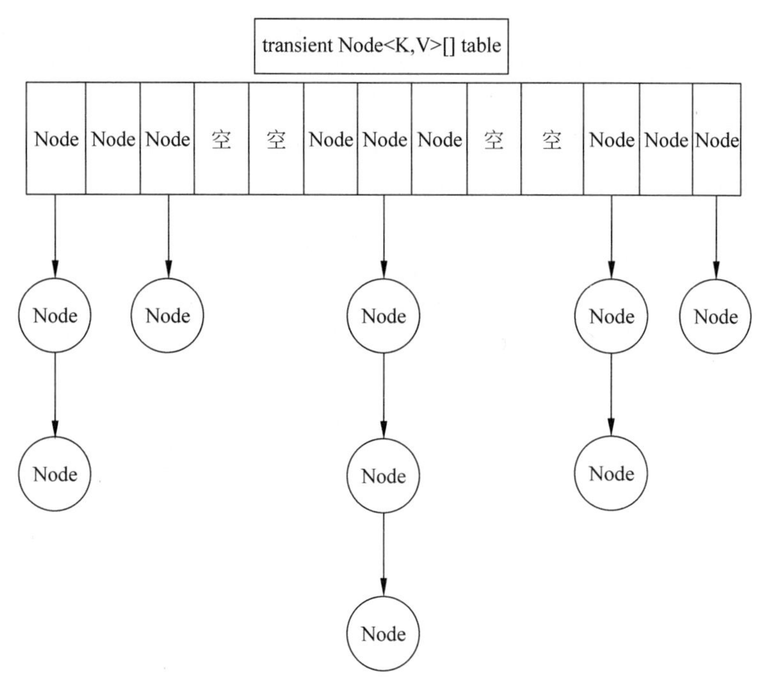

图 16-5 数据存储结构简明示意

表 16-1 HashMap 类构造器

构 造 器	描 述
HashMap()	构造新的对象,默认为无参构造器
HashMap(int initialCapacity)	构造新的对象,指定初始容量
HashMap(int initialCapacity, float loadFactor)	构造新的对象,指定初始容量,指定负载因子
HashMap(Collection <? extends E > c)	构造新的对象,指定初始数据

16.2.3 常用方法

本节基于官方文档介绍常用方法,核心方法配合代码示例以方便读者理解。

1. compute(K key, BiFunction <? super K,? super V,? extends V > remappingFunction)

根据指定的键计算并处理数据,返回计算函数的结果。接收 K 入参,作为指定的键,接收 BiFunction 入参,作为计算函数。

(1) 如果此键不存在并且计算函数的返回结果不为空,则添加键、值数据。

(2) 如果此键存在并且计算函数的返回结果不为空,则替换值数据。

(3) 如果此键存在并且计算函数的返回结果为空,则删除此键、值数据。

代码如下:

```java
//第 16 章/two/HashMapTest.java
public class HashMapTest {

    public static void main(String[] args) {
        HashMap<Integer,String> hashMap = new HashMap<>();
        hashMap.put(156,"one");
        hashMap.put(25,"two");
        //当数据不存在时
        hashMap.compute(33, new BiFunction<Integer, String, String>() {
            @Override
            public String apply(Integer integer, String s) {
                System.out.println(integer + ":" + s);
                return "当数据不存在时";
            }
        });
        System.out.println(hashMap);
        //当数据存在时
        hashMap.compute(33, new BiFunction<Integer, String, String>() {
            @Override
            public String apply(Integer integer, String s) {
                System.out.println(integer + ":" + s);
                return "数据存在-替换";
            }
        });
        System.out.println(hashMap);
        //数据存在,当计算函数返回空时
        hashMap.compute(33, new BiFunction<Integer, String, String>() {
            @Override
            public String apply(Integer integer, String s) {
                System.out.println(integer + ":" + s);
                return null;
            }
        });
        System.out.println(hashMap);
    }
}
```

执行结果如下：

```
33:null
{33=当数据不存在时, 25=two, 156=one}
33:数据不存在时
{33=数据存在-替换, 25=two, 156=one}
33:数据存在-替换
{25=two, 156=one}
```

2. computeIfAbsent(K key, Function<? super K,? extends V> mappingFunction)

如果指定的键关联的值为空，则根据指定的键计算并处理数据，然后返回计算函数的结果，如果指定的键关联的值非空，则返回指定的键关联的值。接收 K 入参，作为指定的键，接收 Function 入参，作为计算函数。

(1) 如果此键不存在并且计算函数的返回结果不为空，则添加键、值数据。

(2) 如果此键存在并且其关联的值为空并且计算函数的返回结果不为空，则替换值数据。

代码如下：

```
//第 16 章/two/HashMapTest.java
public class HashMapTest {

    public static void main(String[] args) {
        HashMap<Integer,String> hashMap = new HashMap<>();
        hashMap.put(156,"one");
        hashMap.put(25,null);
        //当数据不存在时
        hashMap.computeIfAbsent(33, new Function<Integer, String>() {
            @Override
            public String apply(Integer integer) {
                return "数据添加";
            }
        });
        System.out.println(hashMap);
        //如果此键存在并且其关联的值为空并且计算函数的返回结果不为空
        hashMap.computeIfAbsent(25, new Function<Integer, String>() {
            @Override
            public String apply(Integer integer) {
                return "替换数据";
            }
        });
        System.out.println(hashMap);
    }
}
```

执行结果如下：

```
{33 = 数据添加, 25 = null, 156 = one}
{33 = 数据添加, 25 = 替换数据, 156 = one}
```

3. containsKey(Object key)

检查此集合中是否包含指定的键，返回 boolean 值。接收 Object 入参，作为指定的键。

4. containsValue(Object value)

检查此集合中是否包含指定的值，返回 boolean 值。接收 Object 入参，作为指定的值。

5. entrySet()

返回此集合中包含的键、值映射的视图，代码如下：

```
//第 16 章/two/HashMapTest.java
public class HashMapTest {
```

```java
    public static void main(String[] args) {
        HashMap<Integer,String> hashMap = new HashMap<>();
        hashMap.put(156,"one");
        hashMap.put(25,"two");
        Set<Map.Entry<Integer, String>> entries = hashMap.entrySet();
        for (Map.Entry<Integer, String> entry : entries) {
            System.out.println(entry.getKey() + ":" + entry.getValue());
        }
    }
}
```

此方法的底层迭代器实现类是 HashIterator 类，如图 16-6 所示。

```java
abstract class HashIterator {
    Node<K,V> next;           //下一个节点
    Node<K,V> current;        //当前节点
    int expectedModCount;     //防并发修改
    int index;                //当前索引位置
    HashIterator() {
        expectedModCount = modCount;
        Node<K,V>[] t = table;
        current = next = null;
        index = 0;
        if (t != null && size > 0) { //找到第1个节点
            do {} while (index < t.length && (next = t[index++]) == null);
        }
    }

    public final boolean hasNext() {
        return next != null;
    }

    final Node<K,V> nextNode() {
        Node<K,V>[] t;
        Node<K,V> e = next;
        if (modCount != expectedModCount)
            throw new ConcurrentModificationException();
        if (e == null)
            throw new NoSuchElementException();
        //优先匹配节点链路
        if ((next = (current = e).next) == null && (t = table) != null) {
            do {} while (index < t.length && (next = t[index++]) == null);
        }
        return e;
    }
}
```

图 16-6 HashIterator 类的源代码

6. equals(Object o)

对此对象与指定对象进行比较，返回 boolean 值。接收 Object 入参，作为指定对象。

7. get(Object key)

返回指定的键所映射的值，如果此集合中不包含指定的键，则返回空。接收 Object 入参，作为指定的键。

8. getOrDefault(Object key, V defaultValue)

返回指定的键所映射的值,如果此集合中不包含指定的键,则返回默认的值。接收 Object 入参,作为指定的键,接收 V 入参,作为默认的值。

9. isEmpty()

检查此集合是否为空,返回 boolean 值。

10. keySet()

返回此集合中包含键的视图。

11. values()

返回此集合中包含值的视图。

12. put(K key, V value)

将指定的键、值添加到此集合中。返回指定的键相关联的旧值,如果没有,则返回空。接收 K 入参,作为指定的键,接收 V 入参,作为指定的值,代码如下:

```
//第 16 章/two/HashMapTest.java
public class HashMapTest {

    public static void main(String[] args) {
        HashMap<Integer,String> hashMap = new HashMap<>();
        System.out.println(hashMap.put(156, "one"));
        System.out.println(hashMap.put(156, "two"));
        System.out.println(hashMap);
    }
}
```

执行结果如下:

```
null
one
{156 = two}
```

13. putAll(Map<? extends K,? extends V> m)

将指定的集合数据添加到此集合中。接收 Map 入参,作为指定的集合数据。

14. putIfAbsent(K key, V value)

将指定的键、值添加到此集合中。返回指定的键相关联的旧值,如果没有,则返回空。接收 K 入参,作为指定的键,接收 V 入参,作为指定的值,代码如下:

```
//第 16 章/two/HashMapTest.java
public class HashMapTest {

    public static void main(String[] args) {
        HashMap<Integer,String> hashMap = new HashMap<>();
```

```
        System.out.println(hashMap.putIfAbsent(156, "one"));
        System.out.println(hashMap.putIfAbsent(156, "two"));
        System.out.println(hashMap);
    }
}
```

执行结果如下：

```
null
one
{156 = one}
```

注意：如果指定的键已经存在，并且其关联的值也不为空，则此方法不会更新值。

15. remove(Object key)

从此集合中删除指定的键、值数据。返回指定的键相关联的值，如果没有，则为空。接收 Object 入参，作为指定的键，代码如下：

```
//第 16 章/two/HashMapTest.java
public class HashMapTest {

    public static void main(String[] args) {
        HashMap< Integer,String > hashMap = new HashMap<>();
        System.out.println(hashMap.put(156, "one"));
        System.out.println(hashMap.putIfAbsent(256, "two"));
        System.out.println(hashMap.remove(256));
        System.out.println(hashMap);
    }
}
```

执行结果如下：

```
null
null
two
{156 = one}
```

16. remove(Object key, Object value)

仅当指定的键映射到指定的值时，才会从此集合中删除指定的键、值数据，返回 boolean 值。接收 Object 入参，作为指定的键，接收 Object 入参，作为指定的值，代码如下：

```
//第 16 章/two/HashMapTest.java
public class HashMapTest {

    public static void main(String[] args) {
        HashMap< Integer,String > hashMap = new HashMap<>();
        System.out.println(hashMap.put(156, "one"));
```

```
        System.out.println(hashMap.putIfAbsent(256, "two"));
        System.out.println(hashMap.remove(256,"other"));
        System.out.println(hashMap);
    }
}
```

执行结果如下：

```
null
null
false
{256 = two, 156 = one}
```

17. replace(K key, V value)

仅当指定的键存在时替换此键所映射的值。返回指定的键相关联的旧值，如果没有，则为空。接收 K 入参，作为指定的键，接收 V 入参，作为替换的新值。

18. replace(K key, V oldValue, V newValue)

仅当指定的键存在且映射到指定的值时，才替换掉此键所映射的值。返回指定的键相关联的旧值，如果没有，则为空。接收 K 入参，作为指定的键，接收 V 入参，作为指定键的旧值，接收 V 入参，作为替换的新值，代码如下：

```
//第 16 章/two/HashMapTest.java
public class HashMapTest {

    public static void main(String[] args) {
        HashMap < Integer,String > hashMap = new HashMap <>();
        hashMap.put(156, "one");
        hashMap.put(256, "two");
        System.out.println(hashMap.replace(256,"other"));
        System.out.println(hashMap.replace(156,"errorOld","other"));
        System.out.println(hashMap);
    }
}
```

执行结果如下：

```
two
false
{256 = other, 156 = one}
```

19. size()

返回此集合中数据的长度。

16.2.4 TreeNode 类

TreeNode 类继承了 Node 类，当一个索引点链路过长时，可能产生节点类型转换，代码如下：

```
//第 16 章/two/HashMapExtend.java
public class HashMapExtend {

    public static void main(String[ ] args) {
        HashMap<Integer,String> hashMap = new HashMap<>();
        for (int i = 1; i<56; i++) {
            hashMap.put(i * 16 ,"one");
        }
        System.out.println(hashMap);
    }

}
```

Debug 查看数据，如图 16-7 所示。

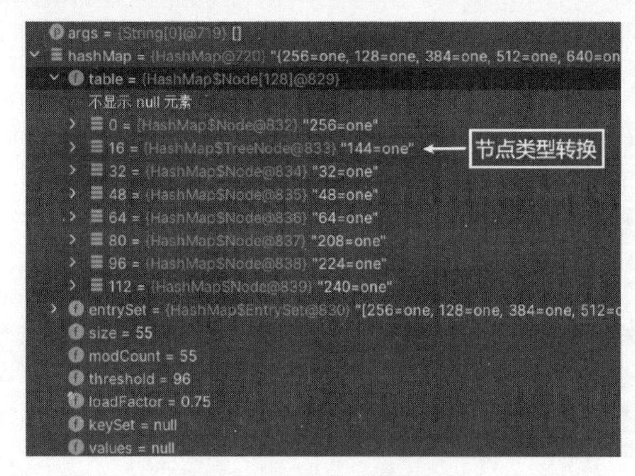

图 16-7 Debug 查看数据

注意：此种节点类型转换是自动完成的，平时在使用的过程中不用刻意关注。在数据量大的情况下这种 TreeNode 节点查询效率比 Node 节点要高。

21min

16.3 LinkedHashMap 类

此类继承了 HashMap 类。此类内部维护了一套链表结构，有两种排序方式，迭代器从链表迭代数据。

16.3.1 数据结构

核心字段如图 16-8 所示。

此类重写 HashMap 类的几个主要方法，如图 16-9 所示。

```
//链表头部
transient LinkedHashMap.Entry<K,V> head;

//链表尾部
transient LinkedHashMap.Entry<K,V> tail;

/**
 * 排序方式
 * 1.访问排序 true
 * 2.插入排序 false
 */
final boolean accessOrder;
```

图 16-8 LinkedHashMap 类的核心字段

```
Node<K,V> newNode(int hash, K key, V value, Node<K,V> e) {
    LinkedHashMap.Entry<K,V> p =
        new LinkedHashMap.Entry<>(hash, key, value, e);
    linkNodeLast(p);//数据链接到链表中
    return p;
}

Node<K,V> replacementNode(Node<K,V> p, Node<K,V> next) {
    LinkedHashMap.Entry<K,V> q = (LinkedHashMap.Entry<K,V>)p;
    LinkedHashMap.Entry<K,V> t =
        new LinkedHashMap.Entry<>(q.hash, q.key, q.value, next);
    transferLinks(q, t);
    return t;
}

TreeNode<K,V> newTreeNode(int hash, K key, V value, Node<K,V> next) {
    TreeNode<K,V> p = new TreeNode<>(hash, key, value, next);
    linkNodeLast(p);//数据链接到链表中
    return p;
}

TreeNode<K,V> replacementTreeNode(Node<K,V> p, Node<K,V> next) {
    LinkedHashMap.Entry<K,V> q = (LinkedHashMap.Entry<K,V>)p;
    TreeNode<K,V> t = new TreeNode<>(q.hash, q.key, q.value, next);
    transferLinks(q, t);
    return t;
}
```

图 16-9 LinkedHashMap 类的主要方法

此类迭代器的实现类是 LinkedHashIterator 类，如图 16-10 所示。

```
abstract class LinkedHashIterator {
    LinkedHashMap.Entry<K,V> next;
    LinkedHashMap.Entry<K,V> current;
    int expectedModCount;

    LinkedHashIterator() {
        next = head; //从头部开始迭代
        expectedModCount = modCount;
        current = null;
    }

    public final boolean hasNext() {
        return next != null;
    }

    final LinkedHashMap.Entry<K,V> nextNode() {
        LinkedHashMap.Entry<K,V> e = next;
        if (modCount != expectedModCount)
            throw new ConcurrentModificationException();
        if (e == null)
            throw new NoSuchElementException();
        current = e;
        next = e.after;
        return e;
    }
}
```

图 16-10 LinkedHashIterator 类的源代码

16.3.2 构造器

LinkedHashMap 类构造器，见表 16-2。

表 16-2 LinkedHashMap 类构造器

构　造　器	描　　述
LinkedHashMap()	构造新的对象，默认为无参构造器
LinkedHashMap(int initialCapacity)	构造新的对象，指定初始容量
LinkedHashMap(int initialCapacity, float loadFactor)	构造新的对象，指定初始容量，指定负载因子
LinkedHashMap(int initialCapacity, float loadFactor, boolean accessOrder)	构造新的对象，指定初始容量，指定负载因子，指定排序方式
LinkedHashMap(Map <? extends K,? extends V> m)	构造新的对象，指定初始数据

16.3.3 常用方法

本节基于官方文档介绍常用方法，核心方法配合代码示例以方便读者理解。

1. compute(K key, BiFunction <? super K,? super V,? extends V> remappingFunction)

根据指定的键计算并处理数据，返回计算函数的结果。接收 K 入参，作为指定的键，接收 BiFunction 入参，作为计算函数。

(1) 如果此键不存在并且计算函数的返回结果不为空，则添加键、值数据。

(2) 如果此键存在并且计算函数的返回结果不为空，则替换值数据。

(3) 如果此键存在并且计算函数的返回结果为空，则删除此键、值数据。

2. computeIfAbsent(K key, Function <? super K,? extends V> mappingFunction)

如果指定的键关联的值为空，则根据指定的键计算并处理数据，然后返回计算函数的结果，如果指定的键关联的值非空，则返回指定的键关联的值。接收 K 入参，作为指定的键，接收 Function 入参，作为计算函数。

(1) 如果此键不存在并且计算函数的返回结果不为空，则添加键、值数据。

(2) 如果此键存在、其关联的值为空并且计算函数的返回结果不为空，则替换值数据。

3. containsKey(Object key)

检查此集合中是否包含指定的键，返回 boolean 值。接收 Object 入参，作为指定的键。

4. containsValue(Object value)

检查此集合中是否包含指定的值，返回 boolean 值。接收 Object 入参，作为指定的值。

5. entrySet()

返回此集合中包含的键、值映射的视图，代码如下：

```java
//第 16 章/three/LinkedHashMapTest.java
public class LinkedHashMapTest {

    public static void main(String[] args) {
        LinkedHashMap<Integer,String> linkedHashMap = new
                                        LinkedHashMap<>(16, 0.75F,false);
            linkedHashMap.put(156,"one");
            linkedHashMap.put(78,"two");
            linkedHashMap.put(455,"three");
            linkedHashMap.put(1545,"four");
            System.out.println(linkedHashMap.get(78));
            System.out.println(linkedHashMap.get(156));
            for (Map.Entry<Integer, String> entry : linkedHashMap.entrySet()) {
                System.out.println(entry.getKey() + ":" + entry.getValue());
            }
    }
}
```

执行结果如下：

```
two
one
156:one
78:two
455:three
1545:four
```

注意：将构造器排序字段修改为 true 并再次运行主方法后观察输出结果。

6. equals(Object o)

对此对象与指定对象进行比较，返回 boolean 值。接收 Object 入参，作为指定对象。

7. get(Object key)

返回指定的键所映射的值，如果此集合中不包含指定的键，则返回空。接收 Object 入参，作为指定的键。

8. getOrDefault(Object key, V defaultValue)

返回指定的键所映射的值，如果此集合中不包含指定的键，则返回默认的值。接收 Object 入参，作为指定的键，接收 V 入参，作为默认的值。

9. isEmpty()

检查此集合是否为空，返回 boolean 值。

10. keySet()

返回此集合中包含键的视图。

11. values()

返回此集合中包含值的视图。

12. put(K key, V value)

将指定的键、值添加到此集合中。返回指定的键相关联的旧值，如果没有，则返回空。接收 K 入参，作为指定的键，接收 V 入参，作为指定的值。

13. putAll(Map <? extends K, ? extends V > m)

将指定的集合数据添加到此集合中。接收 Map 入参，作为指定的集合数据。

14. putIfAbsent(K key, V value)

将指定的键、值添加到此集合中。返回指定的键相关联的旧值，如果没有，则返回空。接收 K 入参，作为指定的键，接收 V 入参，作为指定的值。

15. remove(Object key)

从此集合中删除指定的键、值数据。返回指定的键相关联的值，如果没有，则为空。接收 Object 入参，作为指定的键。

16. remove(Object key, Object value)

仅当指定的键映射到指定的值时，才会从此集合中删除指定的键、值数据，返回 boolean 值。接收 Object 入参，作为指定的键，接收 Object 入参，作为指定的值。

17. replace(K key, V value)

仅当指定的键存在时替换此键所映射的值。返回指定的键相关联的旧值，如果没有，则为空。接收 K 入参，作为指定的键，接收 V 入参，作为替换的新值。

18. replace(K key, V oldValue, V newValue)

仅当指定的键存在且映射到指定的值时，才替换此键所映射的值。返回指定的键相关联的旧值，如果没有，则为空。接收 K 入参，作为指定的键，接收 V 入参，作为指定键的旧值，接收 V 入参，作为替换的新值。

19. size()

返回此集合中数据的长度。

16.4 TreeMap 类

TreeMap 类的底层基于树形节点实现。根据键的自然顺序排序或者根据提供的公共比较器顺序排序。

16.4.1 数据结构

核心字段如图 16-11 所示。

```
//公共比较器
@SuppressWarnings("serial")
private final Comparator<? super K> comparator;

//数据存储
private transient Entry<K,V> root;

//数据长度
private transient int size = 0;

//数据修改标记
private transient int modCount = 0;
```

图 16-11 TreeMap 类的核心字段

数据存储结构简明示意,如图 16-12 所示。

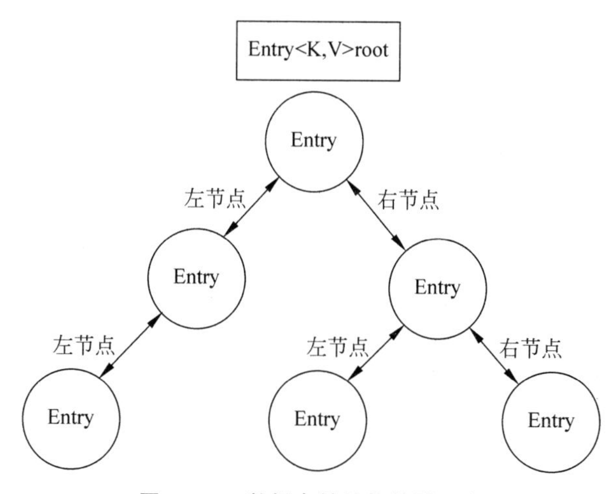

图 16-12 数据存储结构简明示意

注意:排序规则决定了数据是存储在 left 或者 right 链路上。

16.4.2 构造器

TreeMap 类构造器,见表 16-3。

表 16-3 TreeMap 类构造器

构 造 器	描 述
TreeMap()	构造新的对象,默认为无参构造器
TreeMap(Comparator <? super K > comparator)	构造新的对象,指定公共比较器
TreeMap(Map <? extends K,? extends V > m)	构造新的对象,指定初始数据
TreeMap(SortedMap < K,? extends V > m)	构造新的对象,指定初始数据并指定公共比较器

16.4.3 常用方法

本节基于官方文档介绍常用方法，核心方法配合代码示例以方便读者理解。

1. compute(K key, BiFunction <? super K, ? super V, ? extends V> remappingFunction)

根据指定的键计算并处理数据。返回计算函数的结果。接收 K 入参，作为指定的键，接收 BiFunction 入参，作为计算函数。

(1) 如果此键不存在并且计算函数的返回结果不为空，则添加键、值数据。

(2) 如果此键存在并且计算函数的返回结果不为空，则替换值数据。

(3) 如果此键存在并且计算函数的返回结果为空，则删除此键、值数据。

2. computeIfAbsent(K key, Function <? super K, ? extends V> mappingFunction)

如果指定的键关联的值为空，则根据指定的键计算并处理数据，然后返回计算函数的结果，如果指定的键关联的值非空，则返回指定的键关联的值。接收 K 入参，作为指定的键，接收 Function 入参，作为计算函数。

(1) 如果此键不存在并且计算函数的返回结果不为空，则添加键、值数据。

(2) 如果此键存在、其关联的值为空并且计算函数的返回结果不为空，则替换值数据。

3. containsKey(Object key)

检查此集合中是否包含指定的键，返回 boolean 值。接收 Object 入参，作为指定的键。

4. containsValue(Object value)

检查此集合中是否包含指定的值，返回 boolean 值。接收 Object 入参，作为指定的值。

5. entrySet()

返回此集合中包含的键、值映射的视图，代码如下：

```
//第 16 章/four/TreeMapTest.java
public class TreeMapTest {

    public static void main(String[] args) {
        TreeMap<Integer,String> treeMap = new TreeMap<>();
        treeMap.put(3123,"two");
        treeMap.put(343,"three");
        treeMap.put(4545,"four");
        treeMap.put(115,"one");
        treeMap.put(898,"five");
        treeMap.put(458,"six");
        System.out.println(treeMap.entrySet());
    }
}
```

执行结果如下：

```
[115 = one, 343 = three, 458 = six, 898 = five, 3123 = two, 4545 = four]
```

6. equals(Object o)

对此对象与指定对象进行比较，返回 boolean 值。接收 Object 入参，作为指定对象。

7. get(Object key)

返回指定的键所映射的值，如果此集合中不包含指定的键，则返回空。接收 Object 入参，作为指定的键。

8. getOrDefault(Object key, V defaultValue)

返回指定的键所映射的值，如果此集合中不包含指定的键，则返回默认的值。接收 Object 入参，作为指定的键，接收 V 入参，作为默认的值。

9. isEmpty()

检查此集合是否为空，返回 boolean 值。

10. keySet()

返回此集合中包含键的视图。

11. values()

返回此集合中包含值的视图。

12. put(K key, V value)

将指定的键、值添加到此集合中。返回指定的键相关联的旧值，如果没有，则返回空。接收 K 入参，作为指定的键，接收 V 入参，作为指定的值。

13. putAll(Map <? extends K, ? extends V > m)

将指定的集合数据添加到此集合中。接收 Map 入参，作为指定的集合数据。

14. putIfAbsent(K key, V value)

将指定的键、值添加到此集合中。返回指定的键相关联的旧值，如果没有，则返回空。接收 K 入参，作为指定的键，接收 V 入参，作为指定的值。

15. remove(Object key)

从此集合中删除指定的键、值数据。返回指定的键相关联的值，如果没有，则为空。接收 Object 入参，作为指定的键。

16. remove(Object key, Object value)

仅当指定的键映射到指定的值时，才会从此集合中删除指定的键、值数据，返回 boolean 值。接收 Object 入参，作为指定的键，接收 Object 入参，作为指定的值。

17. replace(K key, V value)

仅当指定的键存在时替换此键所映射的值。返回指定的键相关联的旧值，如果没有，则为空。接收 K 入参，作为指定的键，接收 V 入参，作为替换的新值。

18. replace(K key, V oldValue, V newValue)

仅当指定的键存在且映射到指定的值时，才替换此键所映射的值。返回指定的键相关联的旧值，如果没有，则为空。接收 K 入参，作为指定的键，接收 V 入参，作为指定键的旧值，接收 V 入参，作为替换的新值。

19. size()

返回此集合中数据的长度。

20. descendingMap()

返回此集合中数据的反向顺序视图，代码如下：

```
//第 16 章/four/TreeMapTest.java
public class TreeMapTest {

    public static void main(String[] args) {
        TreeMap<Integer,String> treeMap = new TreeMap<>();
        treeMap.put(3123,"two");
        treeMap.put(343,"three");
        treeMap.put(4545,"four");
        treeMap.put(115,"one");
        treeMap.put(898,"five");
        treeMap.put(458,"six");
        System.out.println(treeMap.descendingMap());
        System.out.println(treeMap);
    }
}
```

执行结果如下：

```
{4545 = four, 3123 = two, 898 = five, 458 = six, 343 = three, 115 = one}
{115 = one, 343 = three, 458 = six, 898 = five, 3123 = two, 4545 = four}
```

21. floorEntry(K key)

返回此集合中小于或等于指定数据的最大元素，如果没有此元素，则返回空。接收 K 入参，作为指定数据，代码如下：

```
//第 16 章/four/TreeMapTest.java
public class TreeMapTest {

    public static void main(String[] args) {
        TreeMap<Integer,String> treeMap = new TreeMap<>();
        treeMap.put(3123,"two");
        treeMap.put(343,"three");
        treeMap.put(4545,"four");
        treeMap.put(115,"one");
        treeMap.put(898,"five");
```

```
        treeMap.put(458,"six");
        System.out.println(treeMap.floorEntry(888));
        System.out.println(treeMap.floorEntry(66666));
    }
}
```

执行结果如下：

```
458 = six
4545 = four
```

22. ceilingEntry(K key)

返回此集合中大于或等于指定数据的最小元素，如果没有此元素，则返回空。接收 K 入参，作为指定数据，代码如下：

```
//第 16 章/four/TreeMapTest.java
public class TreeMapTest {

    public static void main(String[] args) {
        TreeMap<Integer,String> treeMap = new TreeMap<>();
        treeMap.put(3123,"two");
        treeMap.put(343,"three");
        treeMap.put(4545,"four");
        treeMap.put(115,"one");
        treeMap.put(898,"five");
        treeMap.put(458,"six");
        System.out.println(treeMap.ceilingEntry(888));
        System.out.println(treeMap.ceilingEntry(66666));
    }
}
```

执行结果如下：

```
898 = five
Null
```

23. subMap(K fromKey, boolean fromInclusive, K toKey, boolean toInclusive)

返回此集合中指定范围内的子元素视图。接收 K 入参，作为开始数据，接收 boolean 入参，作为是否包含开始数据，接收 K 入参，作为结束数据，接收 boolean 入参，作为是否包含结束数据，代码如下：

```
//第 16 章/four/TreeMapTest.java
public class TreeMapTest {

    public static void main(String[] args) {
        TreeMap<Integer,String> treeMap = new TreeMap<>();
        treeMap.put(3123,"two");
        treeMap.put(343,"three");
```

```
        treeMap.put(4545,"four");
        treeMap.put(115,"one");
        treeMap.put(898,"five");
        treeMap.put(458,"six");
        System.out.println(treeMap.subMap(888,true,4545,true));
        System.out.println(treeMap);
    }
}
```

执行结果如下：

```
{898 = five, 3123 = two, 4545 = four}
{115 = one, 343 = three, 458 = six, 898 = five, 3123 = two, 4545 = four}
```

小结

本章节介绍了 Map 接口的常用实现类，读者需要理解各实现类之间的差异。

习题

1. 判断题

(1) HashMap 类的底层基于数组实现。(　　)

(2) LinkedHashMap 类的底层基于链表实现。(　　)

(3) TreeMap 类的底层基于数组实现。(　　)

2. 选择题

(1) 返回 Map 集合数据长度的方法是(　　)。(单选)

A. isEmpty()　　B. keySet()

C. size()　　D. values()

(2) 检查 Map 集合中是否包含指定键的方法是(　　)。(单选)

A. get(Object key)　　B. equals(Object o)

C. containsKey(Object key)　　D. containsValue(Object value)

(3) 检查 Map 集合中是否包含指定值的方法是(　　)。(单选)

A. get(Object key)　　B. equals(Object o)

C. containsKey(Object key)　　D. containsValue(Object value)

(4) 删除 Map 集合中指定键的方法是(　　)。(单选)

A. remove(Object key)　　B. equals(Object o)

C. containsKey(Object key)　　D. containsValue(Object value)

(5) 返回 Map 集合中指定键所映射值的方法是(　　)。(单选)

A. removeAll(Collection <?> c)　　B. contains(Object o)

C. size()　　D. get(Object key)

3. 填空题

查看执行结果并补充代码,代码如下:

```
//第 16 章/answer/OneAnswer.java
public class OneAnswer {

    public static void main(String[] args) {
        LinkedHashMap < Integer,String > linkedHashMap =
                                    new LinkedHashMap <>(16, 0.75F,________);
        linkedHashMap.put(156,"one");
        linkedHashMap.put(78,"two");
        linkedHashMap.put(455,"three");
        linkedHashMap.put(1545,"four");
        System.out.println(linkedHashMap.get(78));
        System.out.println(linkedHashMap.get(156));
        System.out.println(linkedHashMap.get(1545));
        System.out.println(linkedHashMap);
    }
}
```

执行结果如下:

```
two
one
four
{455 = three, 78 = two, 156 = one, 1545 = four}
```

第 17 章

CHAPTER 17

开源 WebSocket 框架

开源 WebSocket 框架基于 Reactor 设计模式，使用原生 NIO 实现的 WebSocket 网络框架，支持多线程、高并发、TLS 安全层协议。

17.1 IM 聊天软件

IM 聊天软件前端基于 HTTP，后端基于 WebSocket 协议实现。

17.1.1 前端展示

前端使用原生的 HTML、CSS、JavaScript 开发。

登录界面，如图 17-1 所示。

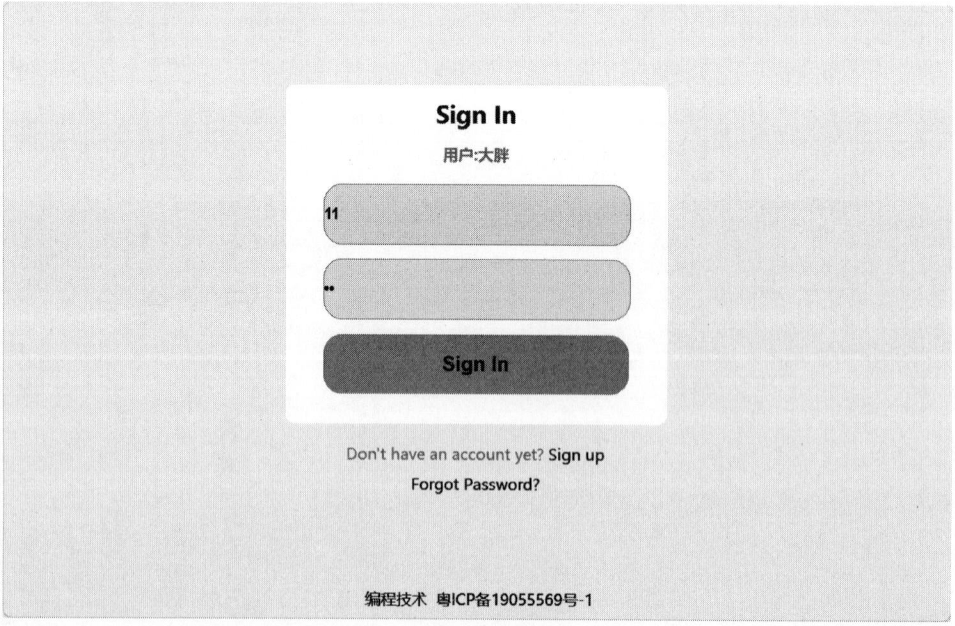

图 17-1 登录界面

添加好友界面，如图 17-2 所示。

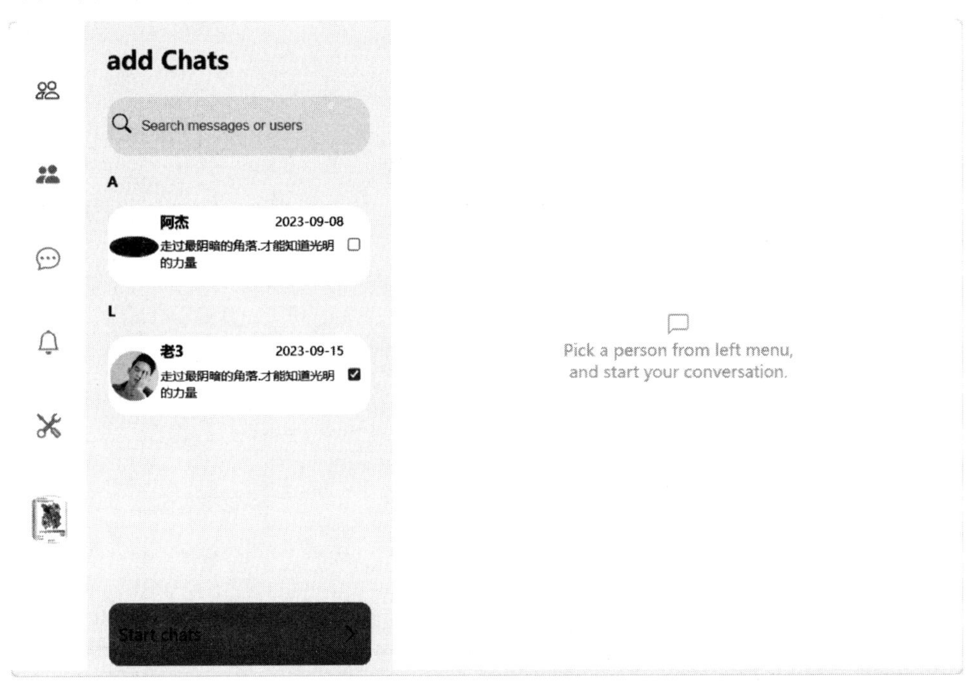

图 17-2　添加好友界面

好友列表界面，如图 17-3 所示。

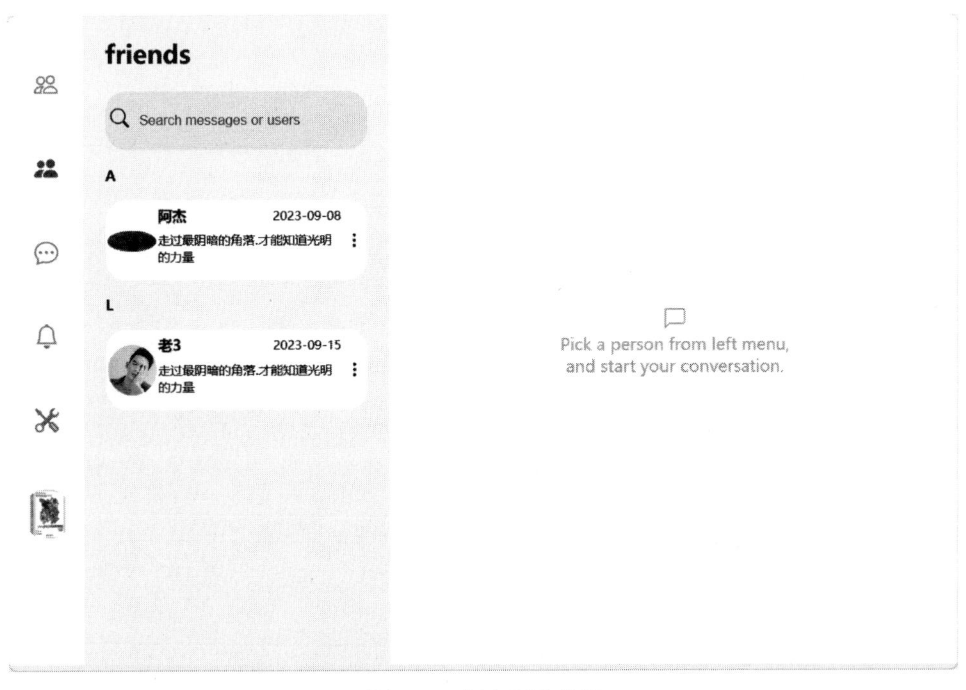

图 17-3　好友列表界面

聊天界面,如图 17-4 所示。

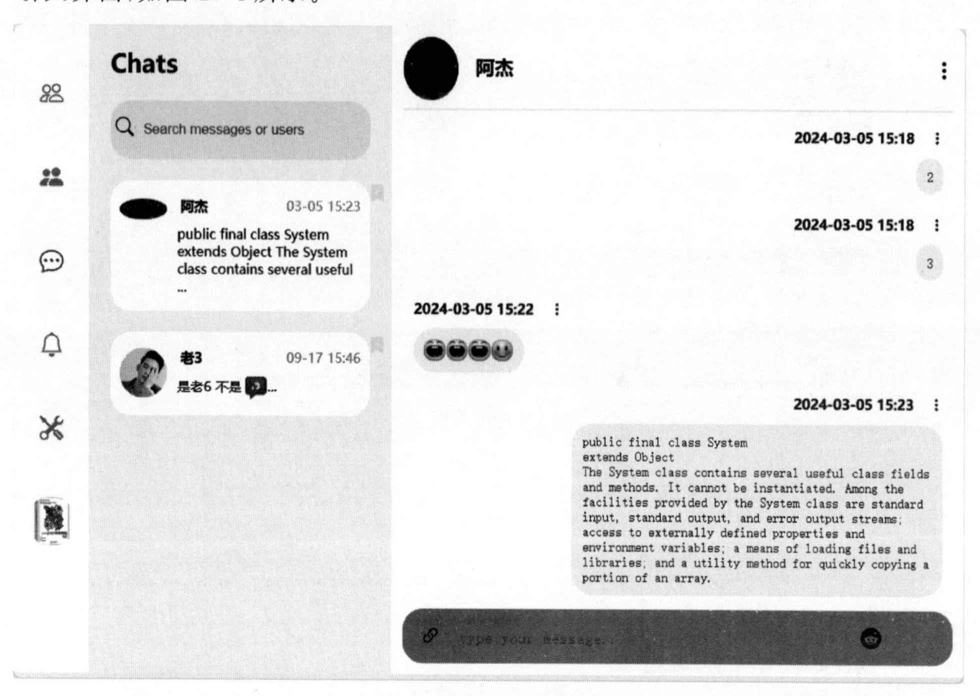

图 17-4　聊天界面

个人设置界面,如图 17-5 所示。

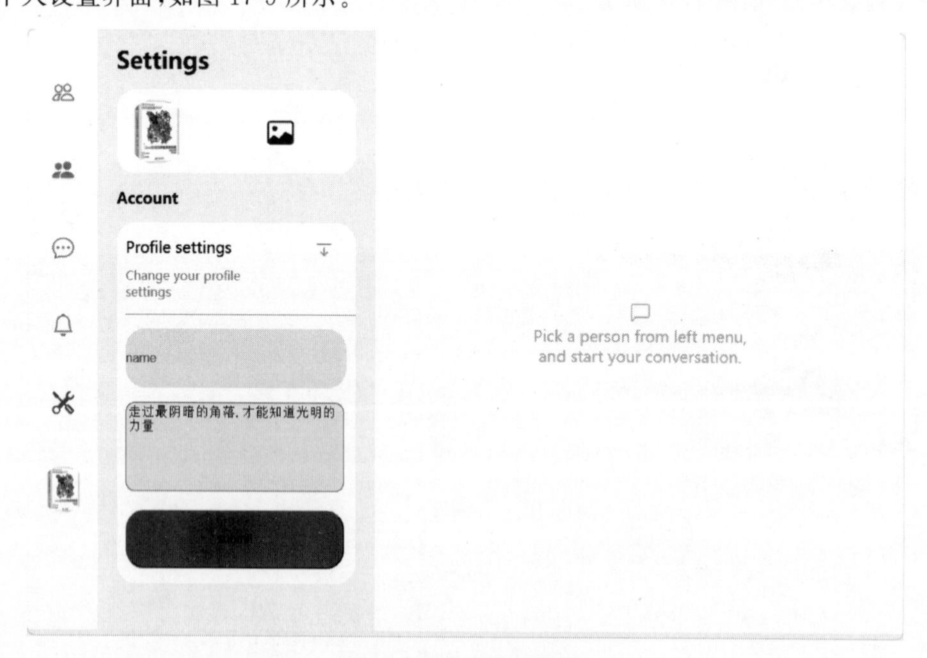

图 17-5　个人设置界面

17.1.2 后端介绍

后端基于Reactor设计模式,使用原生NIO实现的WebSocket网络框架,支持多线程、高并发、TLS安全层协议。第三方依赖MySQL数据库、logback日志框架、Jackson框架、HikariCP数据库连接池。

Maven配置,代码如下:

```xml
<?xml version = "1.0" encoding = "UTF - 8"?>
<project xmlns = "http://maven.apache.org/POM/4.0.0"
    xmlns:xsi = "http://www.w3.org/2001/XMLSchema - instance"
    xsi:schemaLocation = "http://maven.apache.org/POM/4.0.0
                https://maven.apache.org/xsd/maven - 4.0.0.xsd">
    <modelVersion>4.0.0</modelVersion>
    <groupId>cn.kungreat</groupId>
    <artifactId>cpdog</artifactId>
    <version>1.0</version>
    <packaging>jar</packaging>
    <name>cpdog</name>
    <description>nio 框架</description>

    <properties>
        <project.build.sourceEncoding>UTF - 8
                        </project.build.sourceEncoding>
        <project.reporting.outputEncoding>UTF - 8
                        </project.reporting.outputEncoding>
        <maven.test.skip>true</maven.test.skip>
        <skipTests>true</skipTests>
        <java.version>17</java.version>
        <maven.compiler.source>17</maven.compiler.source>
        <maven.compiler.target>17</maven.compiler.target>
        <jackson.version>2.12.4</jackson.version>
    </properties>

    <dependencies>
            <dependency>
                    <groupId>org.projectlombok</groupId>
                    <artifactId>lombok</artifactId>
                    <version>1.18.24</version>
                    <scope>provided</scope>
                    <optional>true</optional>
            </dependency>
            <dependency>
                    <groupId>com.zaxxer</groupId>
                    <artifactId>HikariCP</artifactId>
                    <version>5.0.1</version>
            </dependency>
            <dependency>
```

```
                <groupId>ch.qos.logback</groupId>
                <artifactId>logback-classic</artifactId>
                <version>1.3.0-alpha5</version>
            </dependency>
            <dependency>
                <groupId>mysql</groupId>
                <artifactId>mysql-connector-java</artifactId>
                <version>8.0.22</version>
            </dependency>
            <dependency>
                <groupId>com.fasterxml.jackson.core</groupId>
                <artifactId>jackson-databind</artifactId>
                <version>${jackson.version}</version>
            </dependency>
    </dependencies>

    <build>
        <plugins>
          <plugin>
              <groupId>org.apache.maven.plugins</groupId>
              <artifactId>maven-dependency-plugin</artifactId>
              <version>3.3.0</version>
              <executions>
                <execution>
                <id>copy-dependencies</id>
                <phase>package</phase>
                <goals>
                    <goal>copy-dependencies</goal>
                </goals>
                <configuration>
                <outputDirectory>
                    ${project.build.directory}/lib
                </outputDirectory>
                <overWriteReleases>false</overWriteReleases>
                <overWriteSnapshots>false</overWriteSnapshots>
                <overWriteIfNewer>true</overWriteIfNewer>
                <ExceludeTransitive>false</ExceludeTransitive>
                <ExceludeGroupIds>
                        org.projectlombok
                </ExceludeGroupIds>
            </configuration>
            </execution>
            </executions>
        </plugin>
    <plugin>
    <groupId>org.apache.maven.plugins</groupId>
    <artifactId>maven-jar-plugin</artifactId>
    <version>3.1.2</version>
        <configuration>
          <archive>
            <manifest>
```

```
                <addClasspath>true</addClasspath>
                <classpathPrefix>lib/</classpathPrefix>
                <mainClass>
                    cn.kungreat.boot.CpdogMain
                </mainClass>
              </manifest>
            </archive>
          </configuration>
        </plugin>
      </plugins>
    </build>

</project>
```

源代码类介绍,见表 17-1。

表 17-1　源代码类介绍

类　名　称	类型	描　　述
CpdogMain	class	主程序入口,读取配置文件并初始化必要设置
NioBossServerSocket	interface	套接字服务器端接口类,定义规范方法
NioWorkServerSocket	interface	套接字连接端接口类,定义规范方法
FilterInHandler	interface	过滤器链路接口类,定义规范方法
ConvertDataOutHandler	interface	套接字输出接口类,定义规范方法
ConvertDataInHandler	interface	套接字输入接口类,定义规范方法
ChooseWorkServer	interface	选择套接字连接端接口类,定义规范方法
ChannelProtocolHandler	interface	协议处理器接口类,定义规范方法
CutoverBytes	class	字节数组或整数转换工具类
JdbcUtils	class	数据库连接池工具类
Paging	class	数据库查询分页工具类
WebSocketResponse	class	WebSocket 协议标记位工具类
CpDogSSLContext	class	TLS 安全层协议服务类
InitLinkedList	class	简单的并发安全设计类
ShakeHands	class	线程池类负责 TLS 握手
TLSSocketLink	class	套接字通道数据缓存类
BossThreadGroup	class	自定义线程组实现类
WorkThreadGroup	class	自定义线程组实现类
ChooseWorkServerImpl	class	选择套接字连接端实现类,实现规范方法
NioBossServerSocketImpl	class	套接字服务器端实现类,实现规范方法,接受套接字连接
NioWorkServerSocketImpl	class	套接字连接端实现类,实现规范方法,完成套接字双向通信
WebSocketConvertData	class	套接字输入实现类,实现规范方法
WebSocketConvertDataOut	class	套接字输出实现类,实现规范方法
WebSocketProtocolHandler	class	协议处理器实现类,实现规范方法

续表

类 名 称	类型	描 述
BaseWebSocketFilter	class	过滤器链路实现类，实现规范方法
BaseEvents	class	通知事件处理服务类
ProtocolState	enum	协议标识枚举类
CpdogController	@interface	服务类注解
CpdogEvent	@interface	事件类注解

24min

17.2 WebSocket 协议

WebSocket 协议主要为了解决基于 HTTP/1.x 的 Web 应用无法实现服务器端主动向客户端发送数据的问题，WebSocket 协议可以实现套接字的双向通信。

17.2.1 WebSocket 握手

客户端握手数据，如图 17-6 所示。

```
GET / HTTP/1.1
Host: www.kungreat.cn:9999
Connection: Upgrade
Pragma: no-cache
Cache-Control: no-cache
User-Agent: Mozilla/5.0 (Windows NT 10.0; Win64; x64) AppleWebKit/537.36 (KHTML, like Gecko)
                                                  Chrome/99.0.4844.74 Safari/537.36 Edg/99.0.1150.46
Upgrade: websocket
Origin: https://www.kungreat.cn
Sec-WebSocket-Version: 13
Accept-Encoding: gzip, deflate
Accept-Language: zh-CN,zh;q=0.9,en;q=0.8,en-GB;q=0.7,en-US;q=0.6
Sec-WebSocket-Key: 68/I1/YR1MYqndJnsMN9kQ==
Sec-WebSocket-Extensions: permessage-deflate; client_max_window_bits
```

图 17-6 客户端握手数据

服务器端响应数据，如图 17-7 所示。

服务器端将 Sec-WebSocket-Accept 数据响应给客户端，需要经过一些规范的计算，如图 17-8 所示。

```
HTTP/1.1 101 Switching Protocols
Upgrade: websocket
Connection: Upgrade
Sec-WebSocket-Accept: 1u0599g+GWzsz1THsGQ7X+lpP6c=
```

图 17-7 服务器端响应数据

```java
private String getSecWebSocketKey(String src) throws Exception {
    String rt = src + "258EAFA5-E914-47DA-95CA-C5AB0DC85B11";
    MessageDigest instance = MessageDigest.getInstance("SHA");
    instance.update(rt.getBytes(StandardCharsets.UTF_8));
    byte[] digest = instance.digest();
    return Base64.getEncoder().encodeToString(digest);
}
```

图 17-8 规范的计算

17.2.2 WebSocket 数据交互

握手完成以后就可以进行双向通信了。数据交互规则，如图 17-9 所示。

1字节					1字节		0/2/8字节	4字节	数据
FIN	RSV1	RSV2	RSV3	Opcode	MASK	Payload Len	Extended Payload Len	Masking-key	Data

图 17-9　数据交互规则

数据交互规则详解，见表 17-2。

表 17-2　数据交互规则详解

标　志　位	描　　述
FIN	此标志位用于指示当前的消息是否是一个完整的消息。因为 WebSocket 协议支持将长消息切分为若干数据包发送，切分后的数据只有最后一个数据包为 1，前面的数据包都为 0。如果没有消息切分，则默认为 1
RSV1	保留信息只有在 WebSocket 协议扩展时使用
RSV2	保留信息只有在 WebSocket 协议扩展时使用
RSV3	保留信息只有在 WebSocket 协议扩展时使用
Opcode	0 代表当前是一个延续帧，1 代表当前是一个文本型数据，2 代表当前是一个二进制数据，3～7 为保留帧，8 代表当前是一个关闭帧，9 代表当前是一个 ping 帧，A 代表当前是一个 pong 帧，B～F 为保留帧
MASK	用于指示数据是否使用掩码掩盖，默认为 1，表示使用
Payload Len	以字节为单位指示真实数据的长度，此标志位的长度可变。可能为 7 比特，可能为 7＋16 比特，可能为 7＋64 比特。当此标志位数据为 0～125 时，它的值直接表示数据的真实长度。当此标志位数据为 126 时，其后的 16 比特将被解释为无符号整数，该整数的值表示数据的真实长度。当此标志位数据为 127 时，其后的 64 比特将被解释为无符号整数，该整数的值表示数据的真实长度
Masking-key	掩码掩盖源数据
Data	真实数据

17.3 后端服务

本节介绍后端整体的设计思路、业务思路、数据交互链路。

17.3.1 启动流程

读取配置文件信息并初始化设置，如图 17-10 所示。

主程序入口，如图 17-11 所示。

```
static {
    InputStream cpDogFile = ClassLoader.getSystemResourceAsStream("cpdog.properties");
    Properties props = new Properties();
    try {
        //加载配置文件信息
        props.load(cpDogFile);
        String scanPack = props.getProperty("scan.packages");
        setControllers(scanPack);
    } catch (Exception e) {
        LOGGER.error("CpdogMain-init失败:{}", e.getLocalizedMessage());
    }
    FILE_PATH = props.getProperty("user.imgPath");
    REFRESH_TOKEN_URL = props.getProperty("refresh.token.url");
    //数据库配置
    HikariConfig config = new HikariConfig();
    config.setJdbcUrl(props.getProperty("jdbc.url"));
    config.setUsername(props.getProperty("user.name"));
    config.setPassword(props.getProperty("user.password"));
    config.setAutoCommit(false);
    config.addDataSourceProperty("cachePrepStmts", "true");
    config.addDataSourceProperty("prepStmtCacheSize", "50");
    config.addDataSourceProperty("prepStmtCacheSqlLimit", "512");
    JdbcUtils.dataSource = new HikariDataSource(config);
}
```

图 17-10 读取配置文件信息并初始化设置

```
public static void main(String[] args) throws Exception {
    //创建Boss服务对象
    NioBossServerSocket nioBossServerSocket = NioBossServerSocketImpl.create();
    nioBossServerSocket.buildChannel();
    nioBossServerSocket.buildThread();
    nioBossServerSocket.setOption(StandardSocketOptions.SO_REUSEADDR, true);
    //指定协议处理器
    NioWorkServerSocketImpl.addChannelProtocolHandler(new WebSocketProtocolHandler());
    //指定过滤器链路
    NioWorkServerSocketImpl.addFilterChain(new BaseWebSocketFilter());
    //创建Work服务类数组对象
    NioWorkServerSocket[] workServerSockets = new NioWorkServerSocket[12];
    for (int x = 0; x < workServerSockets.length; x++) {
        //创建Work服务对象
        NioWorkServerSocket workServerSocket = NioWorkServerSocketImpl.create();
        workServerSocket.buildThread();
        workServerSocket.buildSelector();
        workServerSocket.setOption(StandardSocketOptions.SO_KEEPALIVE, true);
        workServerSockets[x] = workServerSocket;
    }
    //创建选择Work服务对象
    ChooseWorkServerImpl chooseWorkServer = new ChooseWorkServerImpl();
    //Boss服务启动
    nioBossServerSocket.start(new InetSocketAddress(InetAddress.getLoopbackAddress(), 9999),
                        workServerSockets, chooseWorkServer);
    //主线程监听事件处理
    GlobalEventListener.loopEvent();
}
```

图 17-11 主程序入口

17.3.2 Boss 服务

NioBossServerSocketImpl 类接受客户端套接字连接,并交给线程池处理后续任务,如图 17-12 所示。

```
@Override
public void run() {
NioBossServerSocketImpl.LOGGER.info("Boss服务启动");
    try {
        while (NioBossServerSocketImpl.this.selector.select() >= 0) {
            Set<SelectionKey> selectionKeys = NioBossServerSocketImpl.this.selector.selectedKeys();
            Iterator<SelectionKey> iterator = selectionKeys.iterator();
            while (iterator.hasNext()) {
                SelectionKey next = iterator.next();
                iterator.remove();
                if (next.isValid() && next.isAcceptable()) {
                    ServerSocketChannel serChannel = (ServerSocketChannel) next.channel();
                    addTask(serChannel);
                } else {
                    NioBossServerSocketImpl.LOGGER.info("Boss事件类型错误");
                }
            }
        }
    } catch (Exception e) {
        NioBossServerSocketImpl.LOGGER.error("Boss线程挂掉", e);
    }
}

private void addTask(ServerSocketChannel serverChannel) throws IOException {
    SocketChannel accept = serverChannel.accept();
    if (accept != null) {
        //交给线程池处理后续任务
        ShakeHands.THREAD_POOL_EXECUTOR.execute(new TLSRunnable(accept));
    }
}
```

图 17-12　NioBossServerSocketImpl 类源代码

17.3.3 TLS 握手

SSLEngine 使用安全层协议 SSL 或 TLS 启用安全层通信。安全层通信流程如图 17-13 所示。

ShakeHands 线程池处理后续任务,如图 17-14 所示。

一个 SocketChannel 套接字通道绑定一个 TLSSocketLink 缓存对象,如图 17-15 所示。

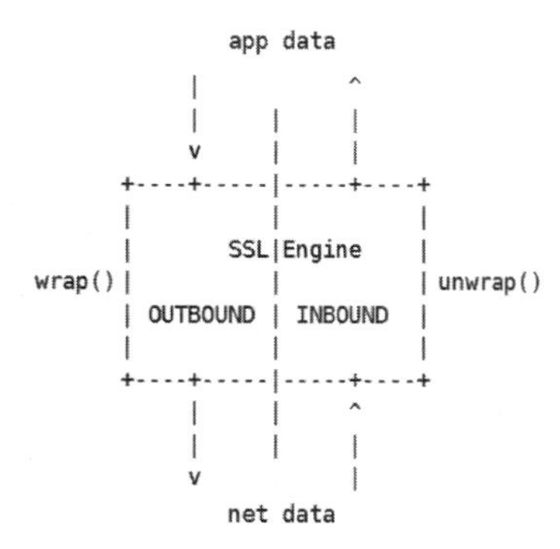

图 17-13　安全层通信流程

17.3.4 Work 服务

NioWorkServerSocketImpl 类监听 OP_READ 事件并完成套接字的双向通信。服务入口如图 17-16 所示。

```
@Override
public void run() {
    TLSSocketLink sslEngine;
    try {
        if (this.tlsSocketChannel.isConnected() || this.tlsSocketChannel.finishConnect()) {
            NioWorkServerSocket choose = NioBossServerSocketImpl.this.chooseWorkServer.choose(
                                        NioBossServerSocketImpl.this.workServerSockets);
            this.tlsSocketChannel.configureBlocking(false);
            choose.setOption(this.tlsSocketChannel);
            //TLS安全层握手
            sslEngine = CpDogSSLContext.getSSLEngine(this.tlsSocketChannel);
            if (sslEngine != null) {
                SelectionKey selectionKey = this.tlsSocketChannel.register(choose.getSelector(),
                                        SelectionKey.OP_READ, sslEngine);
                if (sslEngine.getInSrc().position() > 0) {
                    //说明有多的数据需要处理,添加到Work的初始化队列中去
                    choose.getTlsInitKey().add(selectionKey);
                }
                NioBossServerSocketImpl.LOGGER.info("连接成功{}", this.tlsSocketChannel.getRemoteAddress());
                Thread.State state = choose.getWorkThreads().getState();
                if (state == Thread.State.NEW) {
                    choose.runWorkThread();
                    NioBossServerSocketImpl.LOGGER.info("启动{}", choose.getWorkThreads().getName());
                }
                choose.getSelector().wakeup();
            }
        } else {
            this.tlsSocketChannel.close();
            NioBossServerSocketImpl.LOGGER.info("连接失败{}", this.tlsSocketChannel.getRemoteAddress());
        }
    } catch (Exception e) {...}
}
```

图 17-14　ShakeHands 线程池处理后续任务

```
@Setter
@Getter
public class TLSSocketLink {
    /*
     * 加密解密的引擎
     * */
    private SSLEngine engine;
    /*
     * 从套接字读取的源数据
     * */
    private ByteBuffer inSrc;
    /*
     * 从源数据解密后的数据
     * */
    private ByteBuffer inSrcDecode;
    /*
     * WebSocket 协议完成标识
     * */
    private ProtocolState protocolState = null;

    public TLSSocketLink(SSLEngine engine, ByteBuffer inSrc, ByteBuffer inSrcDecode) {...}

    public void clear() {...}
}
```

图 17-15　TLSSocketLink 类的源代码

```
private final class WorkRunnable implements Runnable {

    @Override
    public void run() {
        //初始化，存出栈的加密数据
        if (CpdogMain.THREAD_LOCAL.get() == null) {
            CpdogMain.THREAD_LOCAL.set(ByteBuffer.allocate(32768));
        }
        try {
            while (NioWorkServerSocketImpl.this.selector.select() >= 0) {
                runTlsInit();
                Set<SelectionKey> selectionKeys = NioWorkServerSocketImpl.this.selector.selectedKeys();
                Iterator<SelectionKey> iterator = selectionKeys.iterator();
                while (iterator.hasNext()) {
                    SelectionKey next = iterator.next();
                    iterator.remove();
                    handler(next);
                }
                clearBuffer();
            }
        } catch (Exception e) {
            NioWorkServerSocketImpl.LOGGER.error(e.getMessage());
        }
    }

}
```

图 17-16 服务入口

数据交互核心流程，如图 17-17 所示。

```
private void baseHandler(SelectionKey next, SocketChannel clientChannel) throws Exception {
    TLSSocketLink attachment = (TLSSocketLink) next.attachment();
    ByteBuffer inSrc = attachment.getInSrc();
    int read = clientChannel.read(inSrc);
    inSrc.flip();
    CpDogSSLContext.inDecode(attachment, 6);
    attachment.getInSrc().compact();
    ByteBuffer inSrcDecode = attachment.getInSrcDecode();
    if (channelProtocolHandler != null && ProtocolState.FINISH != attachment.getProtocolState()) {
        CpdogMain.THREAD_LOCAL.get().clear();
        //WebSocket协议处理
        if (channelProtocolHandler.handlers(clientChannel, inSrcDecode)) {
            attachment.setProtocolState(ProtocolState.FINISH);
        }
    } else {
        List<WebSocketConvertData.WebSocketData> socketDataList = websocketConvert(clientChannel, inSrcDecode);
        if (socketDataList != null) {
            Iterator<WebSocketConvertData.WebSocketData> webSocketDataIterator = socketDataList.iterator();
            while (webSocketDataIterator.hasNext()) {
                WebSocketConvertData.WebSocketData webSocketDataNext = webSocketDataIterator.next();
                if ((webSocketDataNext.getType() == 1 && webSocketDataNext.isDone())
                    || webSocketDataNext.getType() == 8) {
                    runFilterChain(webSocketDataNext, clientChannel);
                    runConvertDataOut(webSocketDataNext, clientChannel);
                    webSocketDataIterator.remove();
                } else if (webSocketDataNext.getType() == 2 && webSocketDataNext.isConvert()) {
                    runFilterChain(webSocketDataNext, clientChannel);
                    runConvertDataOut(webSocketDataNext, clientChannel);
                    if (webSocketDataNext.isDone()) {
                        webSocketDataIterator.remove();
                    }
                }
            }
        }
    }
    if (!clientChannel.isOpen()) {...} else if (read == -1) {...}
}
```

图 17-17 数据交互核心流程

17.3.5 事件服务

GlobalEventListener类负责事件监听、事件响应处理。事件交互核心流程，如图17-18所示。

```
public static void loopEvent() {
    //初始化，存出栈的加密数据
    if (CpdogMain.THREAD_LOCAL.get() == null) {
        CpdogMain.THREAD_LOCAL.set(ByteBuffer.allocate(8192));
    }
    EventBean receiveObj; String receiveObjUrl; SocketChannel socketChannel;
    try {
        while (true) {
            receiveObj = EVENT_BLOCKING_QUEUE.take(); receiveObjUrl = receiveObj.getUrl();
            socketChannel = CONCURRENT_EVENT_MAP.get(receiveObj.getTar());
            if (socketChannel != null && socketChannel.isOpen()) {
                outer: for (int i = 0; i < CpdogMain.EVENTS.size(); i++) {
                Class<?> aClass = CpdogMain.EVENTS.get(i);
                Method[] declaredMethods = aClass.getMethods();
                for (Method methods : declaredMethods) {
                    String name = methods.getName();
                    if (name.equals(receiveObjUrl)) {
                        if (Modifier.isStatic(methods.getModifiers())) {
                            EVENT_BUFFER.clear();
                            CpdogMain.THREAD_LOCAL.get().clear();
                            String rts = (String) methods.invoke(null, receiveObj);
                            if (!rts.isEmpty()) {
                                byte[] bytes = rts.getBytes(StandardCharsets.UTF_8);
                                int readLength = 0;
                                EVENT_BUFFER.put(WebSocketResponse.getBytes(bytes));
                                synchronized (socketChannel) {
                                    do {
                                        int min = Math.min(bytes.length - readLength, EVENT_BUFFER.remaining());
                                        EVENT_BUFFER.put(bytes, readLength, min);
                                        readLength = readLength + min;
                                        EVENT_BUFFER.flip();
                                        CpDogSSLContext.outEncode(socketChannel, EVENT_BUFFER);
                                        EVENT_BUFFER.clear();
                                    } while (readLength < bytes.length);
                                } break outer;
                            }
                        }
                    }
                }
            }
            } else if (socketChannel != null && !socketChannel.isOpen()) {
                CONCURRENT_EVENT_MAP.remove(receiveObj.getTar());
            }
        }
    } catch (Exception e) {
        LOGGER.error("事件监听错误:{}", e.getLocalizedMessage());
    }
}
```

图17-18 事件交互核心流程

小结

本章介绍了 WebSocket 协议握手、WebSocket 协议数据交互。

习题

1. 判断题

(1) TLS 安全层协议握手优先于 WebSocket 协议握手。(　　)

(2) ServerSocketChannel 表示服务器端套接字通道是一个抽象类。(　　)

(3) SocketChannel 表示客户端套接字通道是一个抽象类。(　　)

(4) ServerSocketChannel 只能注册 SelectionKey. OP_ACCEPT 操作集标识。(　　)

2. 选择题

(1) 表示套接字通道的类有(　　)。(多选)

A. ServerSocketChannel 类　　B. SocketChannel 类

C. Socket 类　　D. ServerSocket 类

(2) 设置套接字通道阻塞方式的方法是(　　)。(单选)

A. configureBlocking(boolean block)　　B. isBlocking()

C. validOps()　　D. register(Selector sel, int ops)

(3) 表示套接字通道受支持操作的操作集标识方法是(　　)。(单选)

A. configureBlocking(boolean block)　　B. isBlocking()

C. validOps()　　D. register(Selector sel, int ops)

(4) 表示套接字通道注册到指定的选择器方法是(　　)。(单选)

A. configureBlocking(boolean block)　　B. isBlocking()

C. validOps()　　D. register(Selector sel, int ops)

(5) 表示设置套接字通道指定选项的值方法是(　　)。(单选)

A. configureBlocking(boolean block)

B. setOption(SocketOption < T > name, T value)

C. validOps()

D. register(Selector sel, int ops)

(6) 表示字符集编解码结果的类是(　　)。(单选)

A. CoderResult 类　　B. CharsetEncoder 类

C. CharsetDecoder 类　　D. CodingErrorAction 类

3. 填空题

查看执行结果并补充代码,代码如下:

```
//第 17 章/answer/OneAnswer.java
public class OneAnswer {

    public static void main(String[] args) {
        Charset charset = StandardCharsets.UTF_8;
        CharsetDecoder charsetDecoder = charset.newDecoder();
        ByteBuffer byteBuffer = ByteBuffer.allocate(1024);
        byteBuffer.put("helloWorld-你好世界"
                            .getBytes(StandardCharsets.UTF_8));
        byteBuffer.________________;
        byteBuffer.limit(____________________);
        CharBuffer charBuffer = CharBuffer.allocate(1024);
        CoderResult result = charsetDecoder
                            .decode(byteBuffer,charBuffer,false);
        charBuffer.flip();
        System.out.println(result);
        System.out.println(byteBuffer);
        System.out.println(charBuffer);
        byteBuffer.limit(____________________);
        charBuffer.clear();
        CoderResult resultEnd = charsetDecoder
                            .decode(byteBuffer, charBuffer, true);
        charBuffer.flip();
        System.out.println(resultEnd);
        System.out.println(byteBuffer);
        System.out.println(charBuffer);
    }
}
```

执行结果如下：

```
UNDERFLOW
java.nio.HeapByteBuffer[pos = 20 lim = 22 cap = 1024]
helloWorld-你好世
UNDERFLOW
java.nio.HeapByteBuffer[pos = 23 lim = 23 cap = 1024]
界
```

第18章 虚拟线程

CHAPTER 18

虚拟线程是在JDK 21中正式推出的。可以创建百万级虚拟线程，但不能创建百万级平台线程。虚拟线程的切换成本较低，而平台线程的切换成本较高。

18.1 创建虚拟线程

在JDK 21中Thread类新增加的Thread.Builder API可以创建虚拟线程，代码如下：

```
//第18章/one/VirtualThreadTest.java
public class VirtualThreadTest {

    private static Thread testCurrentThread = null;
    private static final ThreadLocal<String> THREAD_LOCAL =
                                            new InheritableThreadLocal<>();

    public static void main(String[] args) throws Exception {
        THREAD_LOCAL.set("InheritableThreadLocal");
        Thread thread = Thread.ofVirtual().name("VirtualThread-1")
                         .inheritInheritableThreadLocals(true)
                                        .start(new Runnable() {
            @Override
            public void run() {
                System.out.println(Thread.currentThread());
                testCurrentThread = Thread.currentThread();
                System.out.println(THREAD_LOCAL.get());
            }
        });
        Thread.sleep(1000);
        System.out.println(thread == testCurrentThread);
        System.out.println(thread.getName());
        //是否是守护线程
        System.out.println(thread.isDaemon());
        //是否是虚拟线程
        System.out.println(thread.isVirtual());
    }
}
```

执行结果如下：

```
VirtualThread[#30,VirtualThread-1]/runnable@ForkJoinPool-1-worker-1
InheritableThreadLocal
true
VirtualThread-1
true
true
```

在 JDK 21 中 Executors 类新增加的 API 可以创建虚拟线程服务，代码如下：

```
//第 18 章/one/VirtualThreadTest.java
public class VirtualThreadTest {

    public static void main(String[] args) throws Exception {
        try (var executor = Executors.newVirtualThreadPerTaskExecutor()) {
            IntStream.range(0, 10).forEach(i -> {
                executor.submit(() -> {
                    Thread.sleep(Duration.ofSeconds(1));
                    System.out.println(Thread.currentThread());
                    return i;
                });
            });
        }
    }
}
```

执行结果如下：

```
VirtualThread[#37]/runnable@ForkJoinPool-1-worker-9
VirtualThread[#40]/runnable@ForkJoinPool-1-worker-4
VirtualThread[#33]/runnable@ForkJoinPool-1-worker-1
VirtualThread[#30]/runnable@ForkJoinPool-1-worker-8
VirtualThread[#34]/runnable@ForkJoinPool-1-worker-5
VirtualThread[#35]/runnable@ForkJoinPool-1-worker-3
VirtualThread[#32]/runnable@ForkJoinPool-1-worker-2
VirtualThread[#39]/runnable@ForkJoinPool-1-worker-7
VirtualThread[#36]/runnable@ForkJoinPool-1-worker-6
VirtualThread[#38]/runnable@ForkJoinPool-1-worker-7
```

18.2 虚拟线程特点

虚拟线程的主要特点如下：

(1) 虚拟线程并不比平台线程快，并且虚拟线程依赖于平台线程。

(2) 虚拟线程适用于执行大部分时间阻塞的任务，例如等待 I/O 操作完成。虚拟线程不适用于执行长时间计算密集型 CPU 任务。

(3) 虚拟线程使用一组用作承运方线程的平台线程。锁定和 I/O 操作可以将承运方线程从一个虚拟线程切换到另一个虚拟线程。在虚拟线程中执行的代码不知道底层的承运方线程。Thread. currentThread()方法将始终返回虚拟线程对象,即虚拟线程得不到当前的执行线程对象。

(4) 在默认情况下虚拟线程没有线程名称。如果未设置线程名称,则 getName()方法将返回空字符串。

(5) 虚拟线程是守护线程,虚拟线程具有不可更改的固定线程优先级。

(6) 使用 Thread. Builder API 时可以使用 inheritInheritableThreadLocals(boolean inherit)方法来选择是否继承局部变量初始值。

(7) 虚拟线程支持线程局部变量和线程中断。

(8) 虚拟线程的使用并不昂贵,用完即弃。

(9) 调度程序分配虚拟线程的平台线程称为虚拟线程的承运方。虚拟线程可能在其生命周期内安排在不同的承运方上。调度程序不会维护虚拟线程与承运方线程的关联关系。

(10) 承运方线程和虚拟线程的堆栈跟踪是分开的。在虚拟线程中抛出的异常将不包含承运方线程的堆栈信息。线程转储将不会在虚拟线程的堆栈中显示承运方线程的堆栈信息,反之亦然。

(11) 承运方线程的局部变量不可用于虚拟线程,反之亦然。

(12) 虚拟线程的安装和卸载会频繁且透明地发生,并且不会阻塞任何操作系统线程。

(13) JDK 中绝大多数阻塞操作将卸载虚拟线程,并释放其承运方线程以进行新的工作,但是 JDK 中某些阻塞操作不会卸载虚拟线程。如文件系统操作或 JDK 级别 Object. wait()这些阻塞操作的实现通过临时扩展调度程序的并行性来补偿操作系统线程的捕获,因此调度程序 ForkJoinPool 中的平台线程数量可能暂时超过可用处理器的核心数量。可以使用系统属性 jdk. virtualThreadScheduler. maxPoolSize 来调整可用于调度程序的最大平台线程数。

(14) 在两个场景中虚拟线程在阻塞操作期间无法被卸载,因为它被固定到其承运方。在同步代码块或同步方法中执行代码时、执行本机方法或外部函数时。

(15) Thread. getAllStackTraces()现在返回所有平台线程的映射不包括虚所线程。

(16) 虚拟线程不是线程组的活动成员。在虚拟线程上调用 Thread. getThreadGroup()将返回名为 VirtualThreads 的占位符线程组 Thread. Builder API 未定义用于设置虚拟线程组的方法。

18.3 配置承运方线程

可以通过系统属性配置承运方线程,见表 18-1。

表 18-1　配置承运方线程

系统属性	描　述
jdk. virtualThreadScheduler. parallelism	可用于调度虚拟线程的平台线程数，默认为可用处理器核心数
jdk. virtualThreadScheduler. maxPoolSize	可用于调度虚拟线程的最大平台线程数，默认为 256

小结

本章介绍了虚拟线程，读者应理解虚拟线程的使用场景。虚拟线程并不是用来替换平台线程的，两者之间应是相辅相成的关系。

图书推荐

书名	作者
仓颉语言实战(微课视频版)	张磊
仓颉语言核心编程——入门、进阶与实战	徐礼文
仓颉语言程序设计	董昱
仓颉程序设计语言	刘安战
仓颉语言元编程	张磊
仓颉语言极速入门——UI全场景实战	张云波
HarmonyOS移动应用开发(ArkTS版)	刘安战、余雨萍、陈争艳 等
公有云安全实践(AWS版·微课视频版)	陈涛、陈庭暄
虚拟化KVM极速入门	陈涛
虚拟化KVM进阶实践	陈涛
移动GIS开发与应用——基于ArcGIS Maps SDK for Kotlin	董昱
Vue+Spring Boot前后端分离开发实战(第2版·微课视频版)	贾志杰
前端工程化——体系架构与基础建设(微课视频版)	李恒谦
TypeScript框架开发实践(微课视频版)	曾振中
精讲MySQL复杂查询	张方兴
Kubernetes API Server源码分析与扩展开发(微课视频版)	张海龙
编译器之旅——打造自己的编程语言(微课视频版)	于东亮
全栈接口自动化测试实践	胡胜强、单镜石、李睿
Spring Boot+Vue.js+uni-app全栈开发	夏运虎、姚晓峰
Selenium 3自动化测试——从Python基础到框架封装实战(微课视频版)	栗任龙
Unity编辑器开发与拓展	张寿昆
跟我一起学uni-app——从零基础到项目上线(微课视频版)	陈斯佳
Python Streamlit从入门到实战——快速构建机器学习和数据科学Web应用(微课视频版)	王鑫
Java项目实战——深入理解大型互联网企业通用技术(基础篇)	廖志伟
Java项目实战——深入理解大型互联网企业通用技术(进阶篇)	廖志伟
深度探索Vue.js——原理剖析与实战应用	张云鹏
前端三剑客——HTML5+CSS3+JavaScript从入门到实战	贾志杰
剑指大前端全栈工程师	贾志杰、史广、赵东彦
JavaScript修炼之路	张云鹏、戚爱斌
Flink原理深入与编程实战——Scala+Java(微课视频版)	辛立伟
Spark原理深入与编程实战(微课视频版)	辛立伟、张帆、张会娟
PySpark原理深入与编程实战(微课视频版)	辛立伟、辛雨桐
HarmonyOS原子化服务卡片原理与实战	李洋
鸿蒙应用程序开发	董昱
HarmonyOS App开发从0到1	张诏添、李凯杰
Android Runtime源码解析	史宁宁
恶意代码逆向分析基础详解	刘晓阳
网络攻防中的匿名链路设计与实现	杨昌家
深度探索Go语言——对象模型与runtime的原理、特性及应用	封幼林
深入理解Go语言	刘丹冰
Spring Boot 3.0开发实战	李西明、陈立为

续表

书名	作者
全解深度学习——九大核心算法	于浩文
HuggingFace 自然语言处理详解——基于 BERT 中文模型的任务实战	李福林
动手学推荐系统——基于 PyTorch 的算法实现(微课视频版)	於方仁
深度学习——从零基础快速入门到项目实践	文青山
LangChain 与新时代生产力——AI 应用开发之路	陆梦阳、朱剑、孙罗庚、韩中俊
图像识别——深度学习模型理论与实战	于浩文
编程改变生活——用 PySide6/PyQt6 创建 GUI 程序(基础篇·微课视频版)	邢世通
编程改变生活——用 PySide6/PyQt6 创建 GUI 程序(进阶篇·微课视频版)	邢世通
编程改变生活——用 Python 提升你的能力(基础篇·微课视频版)	邢世通
编程改变生活——用 Python 提升你的能力(进阶篇·微课视频版)	邢世通
Python 量化交易实战——使用 vn.py 构建交易系统	欧阳鹏程
Python 从入门到全栈开发	钱超
Python 全栈开发——基础入门	夏正东
Python 全栈开发——高阶编程	夏正东
Python 全栈开发——数据分析	夏正东
Python 编程与科学计算(微课视频版)	李志远、黄化人、姚明菊 等
Python 数据分析实战——从 Excel 轻松入门 Pandas	曾贤志
Python 概率统计	李爽
Python 数据分析从 0 到 1	邓立文、俞心宇、牛瑶
Python 游戏编程项目开发实战	李志远
Java 多线程并发体系实战(微课视频版)	刘宁萌
从数据科学看懂数字化转型——数据如何改变世界	刘通
Dart 语言实战——基于 Flutter 框架的程序开发(第 2 版)	亢少军
Dart 语言实战——基于 Angular 框架的 Web 开发	刘仕文
FFmpeg 入门详解——音视频原理及应用	梅会东
FFmpeg 入门详解——SDK 二次开发与直播美颜原理及应用	梅会东
FFmpeg 入门详解——流媒体直播原理及应用	梅会东
FFmpeg 入门详解——命令行与音视频特效原理及应用	梅会东
FFmpeg 入门详解——音视频流媒体播放器原理及应用	梅会东
FFmpeg 入门详解——视频监控与 ONVIF+GB28181 原理及应用	梅会东
Python 玩转数学问题——轻松学习 NumPy、SciPy 和 Matplotlib	张骞
Pandas 通关实战	黄福星
深入浅出 Power Query M 语言	黄福星
深入浅出 DAX——Excel Power Pivot 和 Power BI 高效数据分析	黄福星
从 Excel 到 Python 数据分析：Pandas、xlwings、openpyxl、Matplotlib 的交互与应用	黄福星
云原生开发实践	高尚衡
云计算管理配置与实战	杨昌家
HarmonyOS 从入门到精通 40 例	戈帅
OpenHarmony 轻量系统从入门到精通 50 例	戈帅
AR Foundation 增强现实开发实战(ARKit 版)	汪祥春
AR Foundation 增强现实开发实战(ARCore 版)	汪祥春